The Great Kitschification

The Great Kitschification

Suzanne Visser

CONTENTS

Foreword

The essays gathered in *The Great Kitschification* are not a screed, nor a lament. They are an aesthetic diagnosis, surgically precise and intellectually generous, of true art versus the simulation of art called kitsch.

Suzanne Visser ties together the sensibilities of artists who could never be mistaken for kitsch—Cy Twombly, Willem de Kooning, Susan Sontag, Marina Abramović, Henri Cartier-Bresson—not to elevate them as icons, but to extract from their work something elemental: the stubborn trace of a human at risk.

A neural net might mimic Twombly's line, but not the force behind it—the stammer, the tremor, the refusal. De Kooning's surfaces, thick with revision, do not perform decisiveness but expose ambivalence in motion. Sontag, even at her most aphoristic, returns the reader to complexity, resisting the closure that kitsch demands. Abramović offers no spectacle for consumption—only the strain of duration and the exposure of the body. Cartier-Bresson's images are not mere timing, but an ethic of waiting—of allowing form to emerge without control.

These are not rhetorical flourishes. They are the ground of Suzanne's claim: that art is not reducible to output, but emerges from a condition of attention no machine can replicate.

This claim is not nostalgic. Suzanne never asks the reader to retreat from technology or to pretend that the past was pure. Instead, she shows how easily an aesthetic of abundance—instant variation, infinite remix—can flood the eye while starving the soul. What is lost, she argues, is not originality per se, but orientation: the slow attunement that links gesture to meaning and maker to world.

These essays do not scold or panic. They invite, with uncommon clarity, a form of looking that resists the quick reward. The defense of authenticity here is not moralistic. It is procedural. She asks: How was this made? By whom? Under what constraints? These questions are not barriers to appreciation, but its preconditions.

Why does this matter now? Because we stand at a moment when the apparatus of imitation is accelerating faster than our ability to discern what is being lost. Suzanne's book does not argue for human superiority. It argues for human specificity. For the value of friction. For the slowness of bodies, the strangeness of hands, the imperfect echo of memory in material. What is offered here is not a doctrine, but a form of resistance—not anti-AI, but pro-art.

These essays arose, in part, from conversations we had while I was working on *The 21st Century Self*. She was the publisher of that book and an attentive reader of its drafts. Her ideas—particularly around authorship, aesthetics, and the implications of artificial intelligence—evolved along with mine, but not in parallel. We are not of one mind on every topic, and that's as it should be. A point of divergence, mentioned here only in passing, concerns the role of machine-generated output in the cultural sphere. Suzanne is cautious, resistant, and rightly concerned with consent and attribution. I remain open to the possibility that AI-generated work may, in some cases, elicit authentic human response. But the difference is not stark, and what unites us matters more: the urgency of reckoning with what AI reveals about the human.

This book is not a polemic. It is a set of inquiries—ranging across painting, photography, language, and labor—into what happens when art becomes indistinguishable from product, and when the conditions of reception become shaped by algorithms and speed. Suzanne does not write from above this dilemma, nor outside of it, but from within. Her position is neither technophobic nor naïve. She is a working artist, immersed in the processes she critiques. What lends the work its force is not opposition to technology, but fidelity to attention. She is not defending a style, but a practice of looking, drawing, living—without automation, without haste, and without the numbing effects of kitsch.

Kitsch, here, is not a cheap insult. It is a structural force: the aestheticization of sentiment, the flattening of ambiguity, the mass production of affect. Suzanne's argument is that kitsch, in its modern form, now includes the slick outputs of generative AI—images that mimic artistic form while bypassing artistic process.

The book's title is not a nostalgic elegy, but a description of what happens when constraint is removed, and consequences elided. In a world where beauty is downloadable and "style transfer" can simulate Twombly, or Jimi Hendrix, or Leonardo Da Vinci, what is left of the human mark?

This question is neither theoretical nor abstract. Suzanne grounds her argument in works that resist this flattening. She returns again and again to the irreducible: the smudge, the pause, the unrepeatable stroke. These are not gestures of genius, but signs of life—of duration, friction, and care. AI can copy the appearance of such marks, but not their origin in time and flesh. That is the core of her insistence: art is not an output, but a trace of attention shaped by a body in the world.

The writing here is lucid, layered, and often lyrical. Suzanne is unafraid to expose her own thinking, to revise, to let ideas breathe. The result is a movement through terrain that feels increasingly unfamiliar: What do we mean by "original" or "authored" in an age of infinite reproduction? What responsibilities do we have to the images we create, and to the humans who encounter them? Can sensitivity survive compression?

Readers of this book will not find final answers, but they will find clarity—earned, not imposed. They will also find an argument for art as an encounter rather than a product.

Suzanne's phrase, "art as an act of tenderness," is not sentimental. It is a refusal to harden to the world. In that, I find myself entirely in agreement.

—Robert Saltzman

1

Introduction

We do not gather in stadiums to watch robots play soccer, or line up for tickets to see a baseball match played entirely by machines. Even if the game were faster, more precise, more statistically perfect than anything human athletes could produce, we would not feel the same shiver, the same pulse of excitement. Why? Because what moves us is not the flawless execution of rules, but the presence of human beings who risk, sweat, fail, recover, and triumph. Sport is not compelling because it demonstrates perfection; it is compelling because it demonstrates failure and humanness. We go to witness our own fragility and possibility mirrored back at us through the imperfect bodies of others.

The same must be said of art. To allow AI to take over art would be as pointless as asking robots to play football. The results might be technically dazzling, perhaps even indistinguishable at first glance from works made by human hands. But what is missing is the very thing that gives sport and indeed any human endeavour its charge: embodiment, intention, failing, pain, vulnerability, and the mark of mortality. A machine cannot know what it means to hesitate before making a stroke, to risk ridicule by showing a painting, or to expose oneself in a poem or other piece of writing.

This is why art must remain the representative of human-ness in the age of AI. In a culture increasingly saturated with algorithmic outputs, polished, repeatable, and perfectly boring, art must stand as testimony to the singularity of human experience. Every risky brushstroke, every faltering line of verse, every hesitant gesture on stage carries with it the entire weight of a lived existence: the childhood that shaped it, the losses and loves that feed it, the accidents that redirect it. Art is not made from patterns; it is made from lives.

The danger we face is not that AI will produce better art than us, but that we will come to accept its kitsch as sufficient. Just as fast food has normalised the replacement of meals cooked slowly and with care, AI threatens to normalise an aesthetic diet of instant imagery and pre-digested clichés. The stadium full of robots kicking a ball may impress with its precision, but it will never elicit the gasp of surprise when a human player bends the laws of physics with an impossible goal. Likewise, AI images may satisfy with their surface polish, but they will never elicit the hush that falls in front of a painting that carries the scars of its maker's life.

To protect this distinction, we must establish legal and cultural frameworks that recognise art as a domain of human meaning, not just of production. Laws must ensure that human cultural heritage is not strip-mined into training data without consent or remuneration. Art education must cultivate the skills of looking, feeling, tasting, smelling and listening as forms of attention that cannot be outsourced. And as audiences, we must learn again to savour what is slow, difficult, and imperfect.

Art, then, should not be defended on the grounds that humans do it "better." That defence will fail, because in many cases the machine will do it faster, slicker, and more abundantly. Instead, art must be defended on the grounds that only humans can do it at all. Just as we watch sport to see humanity embodied in motion, we must turn to art to see humanity embodied in marks, sounds, movements, and words. In an era of proliferating simulacra, art becomes our declaration that we are still here, still feeling, still risking, still alive and still imperfect.

2

A Mirror of Human Banality

The promise of artificial intelligence once conjured visions of machines solving humanity's greatest problems, yet what has emerged instead is a flood of "AI selfies" and digital experiments that feel hollow, decorative, and oddly cursed. Image-generation tools such as DALL-E, Stable Diffusion, and Midjourney have been eagerly adopted as engines for memes, but their outputs have provoked questions about plagiarism, artistic value, and purpose. The results have been more controversial than compelling, ranging from forgettable magazine covers to portraits criticised for their distortions or hypersexualisation.

Curator Lolita Cros, suggest that every art movement is shaped by its medium, and that AI may yet find its place as a common tool. At present, however, the medium overwhelms the message. What captivates is not the art itself but the novelty of the technology. Like Bronzino's portraits, commissioned to flatter rather than to reveal, much AI imagery panders to vanity and cliché. In this sense, the blame may lie less with the machine than with those commissioning it, whose unimaginative demands yield unimaginative results.

Curator Paul Hill, argue that the abundance of bad AI work is simply a continuation of the abundance of bad art throughout history. He sees potential in AI as a resource for sketching ideas rather than replacing human creativity, warning that when people are removed entirely from the process, art itself disappears. What remains vital, he insists, is the narrative: the story of the artist, the struggle, the gesture of life behind the work. Without this, AI imagery feels flat, no matter how polished.

The current wave of AI art is, therefore, not so much a new art form as it is a mirror of human banality. It reveals our appetite for vanity, our impatience for process, and our willingness to confuse novelty with meaning. Like the forgettable portraits of the past, it may one day serve as a reminder that technological possibility is not the same as artistic necessity. Until artists use these tools to embody thought, risk, and message, AI will remain in its childish phase: stick-figure drawings on the refrigerator door of culture. The question is: will this ever change?

3

From the Brain to the Hand 1

We are proud of our children's first drawings. We stick them on the fridge door with magnets, or we frame them and hang them on the wall for everyone to admire. Yet they are not so different from any other child's drawings. Two-year-olds scribble from left to right and back, adding a twirl here and there, a kind of signature. Three-year-olds are fond of circles, presumably depicting the faces of people around them. Eyes, ears and mouths land somewhere on or around the faces and all over the paper. There is fire, and there are dolphins, depending on the child's mind at the moment of drawing. When they are four, children begin to connect the elements. The sun goes in the sky, the house on the ground, the tree next to the house, and the basket in Mum's hand. The connection between the brain and the hand is still wide open. The marks on the paper are fresh. That's why they appeal to us so much.

A child focuses on the process of drawing, not on the result. This changes when the child goes to primary school. At the age of five or six, children begin imitating pictures made by others. They draw what they think they should draw. They imitate their peers and what they see at school and in books. This stage marks the end of good drawing and the beginning of kitsch: the freshness of the marks on the paper disappears.

During the remainder of their childhood, children draw what they learn to draw. The open connection between the brain and the hand is broken. Very few people are aware of this. Most stop drawing when their childhood ends. Only a few keep drawing. Another few begin drawing again during adulthood. Even fewer regain the ability to produce the freshness of line and mark-making they once had.

Most adult amateur drawers produce kitsch. The focus is on the result, not on the process. A tiny fraction of adult drawers focuses on the process, the lines, the marks, and their quality. These are the real artists.

When a child continues to draw into adulthood, she will learn the rules of perspective and foreshortening. The theory of colour comes next, then the use of different media and materials. When all this has been mastered, a "proper" drawing or painting may be the result. What cannot be learned is talent. A talented artist must and will unlearn all that was learned, and start again, from scratch, with a beginner's mind. Everything that has been learned is still there in the background, but this is not what the artist relies upon. This is the point at which great art can begin to emerge.

Some artists chose to go back to the very beginning: they open the gateway between their hand and brain and let nothing they have ever learned interfere. Cy Twombly comes to mind. Others make a conscious decision not to learn or know anything. They are interested in the childhood experience and rekindle it. Once they have re-found it, they remain in the child phase and consciously shut themselves off from any other influences. I have a friend like that. His name is Chris Tate. His drawings are great.

Others are aware of the freshness of the child's lines and blotches, the marks on paper or canvas, and apply the full range of knowledge they have gained. Rembrandt is one example of this kind of painter; another is Frans Hals. In their time, painters did not deviate from the path of technical knowledge. Their response was to set aside all they had learned, reacquire the perspectives of childhood, and then retrieve the skills they had set aside. This gave them the ability to work freely within the constraints of technical knowledge. Their lines and blotches are fresh. They are the greatest masters of all.

Cubism and other abstract movements put an end to the tradition of the 16th-century Dutch masters. Picasso is an obvious example. Painters like Picasso, Karel Appel and Willem De Kooning recognised the importance of the childlike, the naïve, and the primitive in painting. They incorporate these qualities in their work to highlight the limitations of technique and learning.

What is true of drawing and paintings is true of every art form: technique alone is not enough. There must also be talent and the willingness to abandon what one has learned.

Kitsch can easily be recognised, even before it is completed. Kitsch is drawing what you think you see, not what you actually see.

Let's take drawing an eye as an example. Someone who produces kitsch will probably begin a drawing of a figure by drawing the eyes. She has learned that eyes are drawn in a certain way, starting with the eyelids. Then a circle is drawn between them, representing the iris. Into the circle goes a black dot. And then, oh God, no, the eyelashes. What this drawer does not realise is that she cannot see the eyes of the figure she is drawing, let alone the details of the eyes, and let alone the eyelashes. She is too far away. She is not looking; she is "knowing". This makes for a dreadful drawing. And even her "knowing" is not right. The main part of the eye is a ball, which is partly covered by two flaps of skin. The ball is embedded in a hole, the socket, from which it protrudes. All in all, it is a complicated landscape of concave and convex shapes that cannot be taken in at a glance at the beginning of a drawing, when the hand and brain are still impatient.

Starting a figure with the eyes has other drawbacks. Attention might be better given to getting the overall figure onto the page first. What are the dimensions of the body parts? At what angle are they being presented? The imaginary line on which the eyes are located can be indicated at this stage, but not the eyes themselves. They come last, if at all.

Good drawers often draw what is not there. By drawing what is not there, for example, the space around a figure, or the space between an arm and the upper body, the figure emerges on the paper by itself, so to speak. Cézanne threw a sheet over his still-life after he had looked at it for a while. He then painted from memory the empty spaces between the objects. The objects naturally emerged from the space. Michelangelo said that David was already present in the block of marble: his task was to set him free . . . What these great artists have in common is their humility. For them, the artwork was already there, and they merely uncovered it. They did not discover; they uncovered. Good drawers do not need to impose a sense of control. They may have a vague plan for their drawing, but they can abandon it once the hand-brain connection takes over. Lines and blotches appear, and the drawer feels that she is not the one who draws. Drawing is happening. The drawing draws itself. The drawer reveals something that is already there as a possibility.

Decisions made by the drawer come into conscious awareness *after* they have been made. Science has shown this repeatedly since the 1980s, when Benjamin Libet demonstrated it for the first time in his experiments. A good artist knows this instinctively. She does not take credit for the drawing. "The drawing drew itself," she says, or: "God drew it through me" or "my hands drew it."

The drawer of kitsch, on the other hand, has a plan and sticks to it. After the eyes have been put down on paper, other body parts are added one by one in a laborious process towards the imagined end.

The result is constantly borne in mind. A dreadful drawing comes into existence. Every line is dead, and the addition of a huge signature rounds off the entirety. Pity the poor soul who receives it as a gift.

A good drawing or painting is never finished. The drawer decides to stop drawing, or rather, the drawing decides when the time has come to stop. The drawing or painting is still “open”. A kitschy drawing is finished. Very, very finished.

The background in a drawing is as necessary as the subject of the drawing. A good drawer knows this and pays as much attention to it as to any other part of the drawing. A drawer of kitsch treats the background as an afterthought by quickly shading and hatching it. The drawer of kitsch always goes too far and leaves nothing to be explored. The drawing becomes unbearably dense. A monster is created.

The drawer of kitsch is proud. She thinks she has made all the decisions. It all fits together: the blindness to reality, the shading, the huge signature, the pride, the awfulness of the drawing.

AI is the ultimate producer of kitsch imagery. There is no hand and no brain. All elements of the human-made kitsch drawing are present in every image it produces (except, to its credit, the huge signature). There is no seeing, no looking, there is only knowing. And even that knowing is not real knowing.

AI never sleeps. Its visual output is like a tsunami and an avalanche made of kitsch rolled in one, coming at us.

4

Why AI Can Only Produce Kitsch

The rapid advancement of AI into artistic domains has provoked debates over the nature of artistic authenticity and the future of human creativity. AI systems can now generate images, music, text, and film at levels of technical polish that exceed those of human practitioners. Yet these outputs are met with an intuitive but difficult-to-articulate scepticism; they are impressive yet somehow hollow or empty, lacking in depth, boring, uninteresting, exhausting, dead . . . This intuition can be articulated clearly through the theory of kitsch.

Kitsch is not an accidental by-product of AI's so-called art; it is the inevitable outcome of the way such systems operate. The claim advanced here is that AI is structurally incapable of producing art that escapes the domain of kitsch because it lacks being, embodiment, and existential stake.

Human art emerges from lived, messy existence that often engages with messy material such as paint or clay. At the same time, AI generates outputs by recombining statistical patterns drawn from prior works by humans and itself. The absence of ontological grounding ensures that AI's so-called art must remain a simulation rather than an expression.

Clement Greenberg identified kitsch in the early twentieth century as tailored to mass consumption, marked by predictability, sentimentality, and a reliance on familiar formulas that require minimal intellectual or emotional effort to process. Kitsch does not emerge from authentic inquiry but from the mechanical reapplication of established motifs for immediate appeal.

Walter Benjamin's analysis of mechanical reproduction similarly illuminates AI's position in the cultural field. He argued that mass-produced works lose their *aura*, that is, the unique presence and singularity of an artwork situated in a specific time and place. AI-generated works, lacking any originating event in the lived experience of a maker, begin life already stripped of such an aura. They are products with no origin.

Heidegger's conception of art as a mode of *aletheia* (truth-unconcealing) only deepens this analysis. For Heidegger, art emerges from *being-in-the-world*, the situated, embodied existence through which humans encounter and interpret reality. This encounter cannot be reduced to algorithmic processing because it is bound up with temporality, mortality, and the existential stakes of human life. An entity without being, such as an AI system, cannot enter such a mode of truth-disclosure.

Scholars in AI aesthetics have extended these insights, arguing that machine learning models, trained on vast datasets of human cultural output, operate through pattern recognition and recombination rather than intentional creation. As a result, they reproduce statistical regularities that mirror the conformity and sentimentality of human-made kitsch.

Human art is not simply the result of skill in manipulating media; it is a manifestation of lived experience, shaped by the artist's personal history, cultural context, and existential engagement. The artist's embodiment and temporality infuse an artwork with an irreducible singularity. This embeddedness enables art to carry meaning beyond its formal properties, serving as a record of a particular human consciousness encountering the world at a specific moment.

AI's so-called art, by contrast, arises from a process without biography, mortality, pain, celebration, or vulnerability. It is generated through statistical inference to produce new configurations that conform to learned patterns. The AI does not inhabit a cultural horizon; it processes representations of one. It cannot intend, suffer, rejoice, or love; and thus, it cannot create works that bear the trace of such experiences. Its outputs may mimic the external signs of expression, but they lack the internal necessity that characterises authentic artistic work.

Generative AI systems are optimised for outputs that align with user prompts and expectations. This optimisation biases them toward producing works that are recognisable and aesthetically pleasing according to existing conventions. The result parallels kitsch's avoidance of difficulty and pain. AI delivers aesthetic gratification without demanding sustained interpretive engagement. Novelty emerges only within the bounds of pre-established forms. There is no existential risk in its production, no possibility of failure born from an artist's confrontation with the unknown. When instructed to be "original," the system produces permutations of past styles and motifs, not the radical break or subversive gesture that defines avant-garde art.

AI's so-called art is developed and deployed within commercial infrastructures that value scale, speed, and shareability. The platforms that host and distribute AI-generated works incentivise outputs that are instantly appealing and easily commodified. In this respect, AI's so-called art functions as an extension of the market logic that produces kitsch: it is tailored to capture attention in over-saturated cultural marketplaces, optimised for the rapid consumption cycle of digital media.

If AI inevitably produces nauseating kitsch, the challenge for human artists is to make works purely grounded in being. This involves embracing imperfection, risk, and the unpredictable textures of embodied existence. Art that resists kitsch will be that which could not have been made by any other person, in any other moment, for any other reason than the one lived by its maker. Such art cultivates the singularity of the human experience: the flawed brushstroke, the cracked voice, the narrative that emerges from a life lived fully.

A subcategory of kitsch is anecdotal art; works that prioritise telling a literal story over evoking an open, interpretive encounter. In visual art, anecdotalism manifests in paintings that serve as illustrations for a narrative. The wrong medium is used to tell a story. The viewer is guided to decode pre-determined meaning embedded in scenes, characters, and events, rather than being invited into a space of ambiguity or exploration.

Art criticism since the nineteenth century has regarded anecdotal art with suspicion because it reduces art to a vehicle for transmitting a moral, sentimental, or didactic tale. It relies on visual clichés to ensure the story is instantly legible: heroic postures for bravery, bowed heads for grief, and clasped hands for prayer. Such simplification aligns closely with kitsch, which seeks immediate recognition and emotional response at the expense of complexity, subtlety, and the autonomy of form.

Art that shows rather than tells engages the viewer through its composition, melody, texture, rhythm, and the interplay of absence and presence. It does not prescribe a single narrative but opens a space for multiple interpretations, allowing meaning to emerge in the relational encounter between viewer and work.

AI-generated art is predisposed to anecdotalism because its generative process is anchored in language-prompted visual synthesis. Text-to-image systems are trained to produce outputs that match descriptive verbal inputs: "a woman weeping at a graveside in the rain" This prompt-driven mechanism mirrors the illustrative tendency of anecdotal art.

Because AI draws on existing visual patterns to fulfil prompts, it tends to rely on the most statistically prevalent visual signifiers for a given theme; precisely the kind of codified imagery that underpins both anecdotalism and kitsch. The result is a so-called art form that closes rather than opens interpretive space. Viewers are presented with an image that tells them what is happening rather than inviting them to inhabit a world of unresolved possibilities.

The structural reliance on prompts ensures that AI art is and always will be tethered to anecdotal content. Even when an AI system is trained to generate abstract or non-narrative works, the input parameters are still framed by human description, and the model's output remains constrained by the statistical associations of its dataset. It cannot independently initiate the kind of perceptual inquiry or l experiment that produces art which shows rather than tells.

Audience expectations compound this. Much of the fascination with AI's so-called art lies in its capacity to materialise detailed scenes from textual descriptions; a fascination rooted in narrative. AI's so-called art is understood, consumed, and valued as the pictorial realisation of verbal concepts and will remain confined to the anecdotal, and thus to dreadful kitsch.

The implication is that AI is destined to operate within a mode of storytelling that privileges clarity and sentimentality over ambiguity and depth. This is not merely a matter of current technological limitation but of structural logic: a machine without being can illustrate a story but cannot inhabit the perceptual openness from which real art emerges.

Reception theory offers another lens for understanding why AI's anecdotal nature cements its position within the realm of kitsch. Originating in literary studies and later applied to visual art, reception theory shifts the focus from the creator to the audience. It holds that meaning is co-constructed through interpretation rather than transmitted unilaterally from maker to viewer. In non-anecdotal art, this co-construction relies on ambiguity, interpretive openness, and the capacity for a work to sustain multiple, even conflicting, readings. Such art withholds prescriptive meaning and requires the viewer to engage actively, bringing their own experiences, emotions, and cultural knowledge to the encounter. This interpretive labour can generate the kind of transformative engagement that underpins the enduring value of art.

Anecdotal art limits the scope for co-creation by prescribing a singular, literal narrative. The viewer's role is reduced to recognition and emotional confirmation. This recognition operates on a prepackaged set of cultural codes, which short-circuits the process of open-ended interpretation. As a result, the audience's engagement is immediate but shallow, a kind of surface-level gratification instead of a sustained, challenging experience. This dynamic creates a closed feedback loop: the viewer expects the image to conform to a given description, the AI fulfils this expectation, and the successful fulfilment reinforces the expectation itself. Over time, this loop standardises both the production and reception of AI so-called art, precisely the conditions in which kitsch thrives.

From a reception theory perspective, AI's so-called art is thus doubly confined to kitsch alone: first by its anecdotal, illustrative mode of production, and second by an audience engagement model that privileges recognition over interpretation. Without the ontological capacity to produce works grounded in being, and without reception conditions that encourage interpretive depth, AI's so-called art cannot escape being kitsch.

5

From the Brain to the Hand 2

We are proud of the first stories our children tell us. We write them down or even publish them for all to admire. Yet they are not so different from other children's stories. Words merely circulate. There is a limited pool of them in each language. They are not ours. We rearrange them to express ourselves. Some are better at rearranging words than others. This is called talent.

The two-year-old tells incoherent stories. Her mind jumps. Her hands move....

The three-year-old is fond of telling stories about people and things around them: Mum and Dad, fire, dolphins and dinosaurs, depending on what is in the child's mind when it is telling the story. Around the age of four, children begin to connect the elements. The sun goes in the sky, the house sits on the ground, the tree is next to the house, the basket is in Mum's hand, the dolphin lives in the water, and the dinosaur eats everything. The connection between the brain and the hand is still wide open. The words in the story are fresh. That's why they appeal to us.

The child focuses on the process of telling a story, not on the result. This changes when the child goes to primary school. At the age of five or six, children begin to tell the stories of others. They pick them up from their environment, their peers, and what they learn at school and read in books. This is the end of good storytelling and the beginning of kitsch. The freshness of the words in the story disappears.

During the rest of childhood, children relate stories in ways they learn from other people. The open connections between the brain and the mouth and hand are broken.

Very few people are aware of this. Most stop telling stories when childhood ends. A few keep telling stories. A few begin telling or writing stories again during adulthood. Even fewer regain the ability to produce words with the freshness they once had.

Most adult amateur writers produce kitsch. The focus is on the result, not on the process. AI is in this category.

A tiny fraction of humanity focuses on the process, the words, the sentences, the paragraphs, and their quality. These are the real artists.

Words appear, just like thoughts. Nobody has any idea how. I have no idea what my next sentence is going to be when I write this.

When a child continues to tell stories, she will learn the rules of spelling and grammar. Literature is next. Then, the use of different styles of writing. A proper book may result when all this has been mastered. What cannot be learned is talent. Talent is either there or it is not. It is given by nature.

Some make a conscious decision not to learn or know anything. They are interested in the childhood experience and rekindle it. Once they have found it, they remain in the childlike phase and knowingly close themselves off from any influences. These are the writers who write children's books as if they were written by children, or artists like Laurie Anderson and Yoko Ono.

Magical Realism and other abstract movements put an end to the tradition of the master writer. Children's authors recognised the importance of the naïve and the primitive, the expressiveness of these styles. They incorporated these in their work to point out the restrictions of technique and learning.

What is true of writing is true of every art form: technique alone is not enough. There must also be talent and the willingness to abandon what one has learned. Sticking to the rules of grammar and spelling does not make literature. AI has neither talent nor a willingness to abandon what it has learned. It sticks to the rules no matter what. Kitsch can easily be recognised even before it is made. Kitsch is writing about what you think you see and not writing about what you see. AI has no eyes. It cannot see.

Let's take an eye as an example: Someone (including AI) who produces kitsch will probably begin describing a character by describing the eyes. She has learned that eyes are described in a certain way. First the colour, then the mood, then, oh God, no, the story the eyes tell. What she does not realise is that she most probably cannot see the eyes of the figure she is describing, let alone the details of the eyes, let alone the colour. She is not looking; she is "knowing". This makes for a dreadful description. And even the "knowing" is not quite right. The central part of the eye is a ball which is partly covered by two flaps of skin. The ball is embedded in a hole, the socket, from which it protrudes. All in all, it is a complicated landscape of concave and convex shapes that cannot be taken in at a glance at the beginning of a story, when the hand and brain are still impatient. AI has no brain or hands that are racing ahead.

Describing a character's eyes can have other drawbacks.

Attention might be better given to getting the whole character onto the page first. What are the main characteristics? At what angles do they present themselves? The eyes can be indicated at this stage.

The worst book has the showiest cover. The obscure name of the writer screams at us: *Belinda Castle*. (Who the F is Belinda Castle?) John Grisham is a writer like this. AI is a writer like this.

Good writers do not need to impose control. They may have a vague plan for the story, but they can abandon it once the hand-brain connection takes over. Words and sentences appear, writing is happening, and the writer feels that she is not the one who is writing. The story tells itself. The writer reveals something that is already there as potential. The writer is not discovering something but uncovering it. Robert Saltzman comes to mind.

Decisions made by the writer come into conscious awareness *after* they have been made. Science has shown this repeatedly since the 1980s, when Benjamin Libet demonstrated it for the first time. A good artist knows this instinctively. She does not take credit for the story. "The story wrote itself," she says, or "God wrote it through me."

The author of kitsch, including AI, on the other hand, has a plan and sticks to it. After the eyes have been described, other parts are added in a laborious process towards the imagined end. We learn of the hair colour, the scarf, and the handbag. The end is constantly borne in mind. A dreadful story is brought into existence. Its words and sentences are dead; its characters have no volume. A loud cover rounds it all off. Pity the poor soul who receives it as a gift!

A good story or book is never finished. The author of true literature decides to stop writing, or rather, the story decides when the time has come to stop. The story is still "open" to changes and remains that way. A kitsch story is finished. Very, very finished. AI writes the ultimate kitsch story.

The background in a story is as necessary as the subject of the story. A good author knows this and pays as much attention to it as to any other part of the story. The characters don't only have a front, but also a back. Some have a tail.

A writer of kitsch treats the background as an afterthought. Her characters are flat, cardboard cut-outs. This is AI in a nutshell. A writer of kitsch uses clichés. AI only uses clichés. The writer of kitsch, including AI, always goes too far in editing and leaves nothing unsaid. The story becomes unbearably dense. A monster is created.

The writer of kitsch is proud. She thinks she has made all the decisions. It all fits together: the blindness to reality, the shading, the cardboard characters, the clichés, the loud cover, the pride, and the awfulness of the language.

This is the signature of AI when it tries to produce art.

6

Suffering and Mortality

Art has always been bound to the conditions of human finitude. Artistic expression carries within it the marks of impermanence, grief, fragility and mortality. These conditions are not accidental to art, but constitutive of its very possibility. To create is to know that the creation and the creator alike are destined to perish, and that what survives may only ever be a fragment or echo of what was once whole.

Suffering has often provided the material from which art arises. Personal grief finds form in lamentation, social struggle is recorded in song or mural, and historical trauma becomes visible in literature and film. The work of art transforms anguish into shape, allowing it to be witnessed and shared. It is through this process that suffering, though deeply private, enters the realm of collective giving of meaning. The weight of experience finds transfiguration in colour, word, sound or gesture. This is not a matter of representation alone. It is the alchemy of turning pain into form, an act that preserves vulnerability while offering it to others as recognition and witness.

Mortality, likewise, infuses art with significance. Every brushstroke, every line of verse, carries within it the knowledge that both artist and audience are finite beings. The painter who layers oil on canvas knows that the canvas will one day deteriorate, and that their own body will cease to be. The composer who arranges notes into a requiem acknowledges both loss and the inevitability of further loss. Comic artwork gains depth against the backdrop of death, for the affirmation of life is intensified by the awareness of its limits. Without mortality, art would lack the tension that gives it gravity.

Artificial intelligence cannot enter into this domain. It can simulate the surfaces of grief and death by recombining visual tropes or narrative motifs, but it cannot suffer and it cannot die. The system is indifferent to the content it produces. AI cannot feel the shudder of loss, nor can it sense the shadow of its own end. It remains untouched by the existential conditions that shape human creation.

The absence of suffering and mortality renders AI incapable of producing art that bears existential weight. Its works are decorative rather than revelatory, spectacle rather than testimony.

The distinction is decisive. Art arises from the vulnerability of the maker. AI is not vulnerable; it has no life to lose. It does not stand on the edge of death, nor does it wrest meaning from pain. For this reason, it can only reproduce kitsch: ornaments of mortality that carry none of its truth.

If art is inseparable from the human confrontation with suffering and death, then AI will never create art. It will only create representations emptied of the very conditions that give art its necessity. To mistake these representations for art is to mistake imitation for being, and spectacle for truth.

7

Clichés and Stereotypes

Research by The Conversation reveals that when artificial intelligence turns its gaze toward, say, Australia, it does not see the living, layered reality of the place. It sees fragments, patterns stripped from their roots, shapes and colours severed from the histories that give them meaning. The bush becomes a generic backdrop of gum trees and ochre skies. Aboriginal culture becomes a collection of inadequately represented dot paintings, boomerangs, black silhouettes against a sunset and, oh give us a fucking break, huts. These images are not drawn from relationship, knowledge, or lived experience. They are scraped from the digital commons, flattened into clichés and stereotypes, and returned to us as though they were representations of who we are. The vast complexity of the Australian continent is reduced to a handful of visual tropes that can be easily recognised and reproduced. The coastal cities are rendered as tourist-brochure skylines. The deserts are empty expanses, devoid of the human and ecological relationships that have shaped them for tens of thousands of years. Indigenous people, if depicted at all, are stylised into a nauseating shorthand designed for consumption by outsiders. This is not the Australia lived by its people; it is a pastiche assembled from what the algorithm has found most statistically "Australian" in the sea of data it has ingested.

The result is a new kind of mythmaking. But unlike the old myths, which at least arose from human imagination, this myth is a mechanical recycling of what is most available and most repeated. And because AI systems are trained to optimise for recognisability rather than truth, they privilege the familiar stereotype over a more unfamiliar reality. The machine not only reflects but amplifies the biases embedded in its training material. The more we feed it the same clichés, the more confidently it will reproduce them, until the feedback loop is complete and no alternative images of the nation survive in the cultural bloodstream.

This is not a harmless distortion. It is a continuation of a pattern in which Indigenous culture has been extracted, decontextualised, and commodified for the consumption of others. In human hands, this has taken the form of souvenir-shop appropriations or corporate branding campaigns. In machine hands, the process is faster, vaster, and more insidious. An algorithm trained on a shallow diet of images will reproduce the same narrow tropes over and over, because it has no capacity to understand. It reinforces stereotypes, erases complexity, and drowns out the voices of those who live and shape the cultures it depicts.

Just as the appropriation of Indigenous art styles into mass-market products strips them of context and meaning, AI-generated Australiana strips the whole country of its human, ecological, and historical depth. In the end, we are left with clichés and stereotypes; familiar-looking shells with their inner life removed.

There is another danger lurking. Once such shallow representations saturate the cultural environment, they begin to overwrite memory itself. A child growing up in an AI-mediated visual world will learn to recognise Australia not as it is experienced, but as it appears in machine-generated images. AI becomes a co-author of the national imaginary, but one without conscience, without curiosity, and without the capacity to be corrected by reality.

Resisting this requires more than just better datasets or more inclusive prompts. It requires the assertion of cultural sovereignty by all Australians. Consent, attribution, and fair benefit-sharing are not optional extras. They are the conditions under which cultural material can be used without becoming a fresh act of dispossession. Without them, AI-generated Australiana will continue to be a mirror that reflects only the stereotypes and cliches already lodged in its training data, never the truth.

The alternative is still possible. It begins with recognising that the authority to depict a culture belongs to those who live it. It involves building systems that are accountable to the communities whose images and stories they use. And it means insisting that art retains its function as a vessel for tenderness, truth, and relationship. Without it, the machine will remain what it currently is: an engine for mass-producing stereotypes and cliches, sold back to us as though they were our reflection.

The theft is compounded by its invisibility, masked by the cheerful familiarity of the output. This is the same machine that now turns on all of us, mining the creative commons of humanity, rebranding it as novelty, and selling it back to us as if it were our reflection. The archives of human culture, built across centuries, are being emptied into vast machine learning systems that recognise no context and owe no allegiance to the past. The continuity between creation and creator is broken; the temporal thread is cut. This is not just theft of content; it is theft of the time that made the content possible.

8

Perfect Copies, Perfectly Boring

Out of curiosity, I once commissioned paintings from a workshop in China that specialised in reproducing famous artworks on demand. I ordered a Malevich and a Mondrian. Both pieces I knew intimately from years of visiting the Stedelijk Museum in Amsterdam. I had stood before them many times, with my father, who was a painter and taught me about art from when I was very young.

The Chinese copies arrived after several months. They were flawless, at least by the standards of visual fidelity. Every colour block and line matched my memory of the originals. Yet as soon as I had unwrapped them, I felt a strange combination of disappointment, shame, irritation, and hatred. The works were identical and, at the same time, dead, dull, stupid, and somehow extremely ugly. The very perfection of their reproduction seemed to strip away the qualities that made the originals so compelling.

The failure was not technical. The Chinese copy painters had fulfilled the task exactly as commissioned. Their work was a triumph of manual precision. What was missing was something the commission itself had caused: the original act of creation, the historical context, and the artistic intention and integrity that had given the works their meaning.

When Mondrian painted *Composition with Red, Blue and Yellow*, it was not merely an arrangement of colours within black lines; it was a radical simplification of form that carried the weight of years of experimentation and philosophical reflection on abstraction.

Malevich's *Black Square* was not simply that, a black square; it was a manifesto in paint, a rupture with representational art, an emblem of Suprematism. The Chinese handmade copies could match the physical arrangement of pigment, but not the conceptual act that made those arrangements so significant.

The result was something akin to taxidermy; an image of life without life itself. The surface details were there, but the work's presence, its *aura*, to borrow Walter Benjamin's term, was absent.

The relationship between those painted reproductions and AI-generated art is direct. Generative AI systems, like the Chinese copy painters, excel at replication. They can reproduce styles, compositions, and even imperfections. Yet AI so-called art, like the commissioned copies, operates without access to the origin point of meaning. It lacks the historical moment, the existential stakes, and the artist's embodied engagement with materials. It can *do,* but it cannot *be*. This is why, despite possible fake virtuosity, AI art provokes in critical viewers the same reaction I had to my Malevich and Mondrian copies: boredom, irritation, disappointment, disgust, and even a sense of hatred.

Perfection in replication is always aesthetically deadening because it eliminates the unpredictability that comes with the act of making. In a human-made artwork, the slightest deviation from an ideal, whether deliberate or accidental, can carry enormous expressive weight.

In AI's so-called art, as in a copied painting, the goal is to eliminate deviation. The system is trained to approximate an idealised representation as closely as possible. In doing so, it produces images that are smooth to the point of death by sterility.

The ugliness of such work is not merely a matter of taste. It emerges from its sheer banality; the fact that, stripped of historical, cultural, and personal context, all that remains is a configuration of forms that can be endlessly and meaninglessly repeated.

If Chinese copy painters can reproduce a Malevich or a Mondrian without contributing anything to the history of art, AI magnifies this condition exponentially. But instead of a few commissioned reproductions, AI can flood the entire cultural field with billions of stylistically perfect yet conceptually empty images in minutes. Such saturation leads to what might be called *kitsch tsunami nausea*: the sense that we are drowning in images that lack the raw necessity of art. The banality lies in the overwhelming abundance. Like perfect copies stacked floor to ceiling in trillions of warehouses, they become oppressive through their sheer interchangeability.

Perfect replication is not neutral; it is reductive. Whether by hand, camera, scanner, or algorithm, the act of copying flattens and hollows the multidimensional life of an artwork into a single plane of appearance. The result becomes increasingly disturbing as the number of copies grows.

The banality of AI's so-called art and its endless production may carry within it an unintended gift. Saturation has a way of sharpening perception. Just as a glut of processed food can awaken a craving for something organically grown in soil, an oversupply of dead, machine-made imagery may deepen the appetite for works that carry the unpredictable, risk-and pain-laden marks of human making. Live performance, site-specific installation, and works embedded in cultural traditions may become valued precisely because they cannot be copied. Over time, the cultural nausea induced by AI's aesthetic banality may well push audiences back toward those living works that are messy, smelly, flawed, embodied and therefore alive.

9

Pierre et Gilles

Pierre Commoy and Gilles Blanchard are working together since the late 1970s under the name *Pierre et Gilles.* They make staged portraits that straddle painting, photography, and digital retouching. Their signature style involves photographing sitters, often celebrities, models, or queer icons, against elaborate sets, then hand-painting the prints with glowing, decorative details. The result is a hyper-saturated aesthetic that oscillates between camp glamour and devotional iconography.

Their art is unashamedly hyper-kitsch. The theatrical staging, exaggerated sentimentality, and artificial perfection recall religious paintings, Bollywood posters, and advertising photography, but always with a wink. By collapsing distinctions between the sacred and the profane, high art and pop culture, they interrogate what happens when taste is overwhelmed by spectacle.

For Pierre et Gilles, kitsch is not a by-product but a strategy. Their portraits revel in surfaces, ornament, and fantasy, asking viewers to suspend judgment about authenticity. Figures like Madonna, Boy George, and handsome sailors are transformed into icons that shimmer with a sense of artificial paradise. The result is both seductive and discomforting: beauty taken to an extreme, where it blurs into parody.

This approach mirrors AI-generated kitsch, which overwhelms viewers with its sheer surface spectacle: glossy perfection, endless repetition, and emotional exaggeration. Just as Pierre et Gilles craft an idealised version of their sitters, AI algorithms tend to idealise their subjects into flattened archetypes, producing images that feel at once familiar and hyper-artificial, but without the wink.

Pierre et Gilles highlight how kitsch intersects with queer culture. Their work embraces artifice, theatricality, and excess as modes of resistance against heteronormative taste.

AI systems can reproduce and amplify these aesthetics of camp and glamour, but strip them of their political and cultural context. What emerges is a kind of algorithmic queer kitsch, where stylistic markers survive but their critical charge disappears.

If Koons demonstrates kitsch as a capitalist celebration and Murakami turns it into a Superflat consumer spectacle, Pierre et Gilles exemplify kitsch as synthetic iconography.

Pierre et Gilles prefigure AI's tendency to generate endless "beautiful surfaces". But the AI-generated product is devoid of historical anchoring.

10

Looking and Seeing

To be a skilled human drawer or painter, one must learn how to see truly. The act of looking and seeing is an art form in itself. If painting or drawing is not the result of observation, but rather imagination alone, it can easily become kitsch. In my opinion, Salvador Dalí is the ultimate master of kitsch, as his art is entirely born from his imagination without any actual observation of the world around him. I know this is called surrealism, but that doesn't make it better. People who are fond of Dalí often have specific other characteristics; they may also enjoy Pink Floyd, believe in conspiracy theories, and are fascinated by UFOs. The reason for this combination remains a mystery to me, but it is telling at the same time.

The process of looking and seeing is a complex one. For example, if I cut a shape out of a white piece of paper and place it back onto the same white paper, I will not see the shape. However, if I crumple the shape a little, some parts of it will float slightly off the paper, and I will see the shape. This is because the shadow beneath the shape gives it definition against the background.

When we observe things, we are always looking at two things at once: the object and its background. The difference between the figure and the background gives the figure its identity. We can only experience an object or figure by comparing it with something else.

Thus, seeing and looking are essentially acts of distinguishing. It is impossible to see or look at the essence of a thing. We never see things directly.

The passage of time also plays a role in what we see when we look. When we observe a nearby object, we see it in almost real-time, even though it takes a fraction of a second for the image to travel from our eyes to our brain. When we observe objects that are very far away, such as stars, we see a version of them that existed a long time ago. In this way, we look into the past when we observe the universe. Some objects that we see in the sky do not exist any longer.

Looking at things very closely, such as through a microscope, results in an even more bewildering experience. We encounter a world that is entirely unknown to us.

Closer to home, in our daily lives, there must be a relationship between two things before we can perceive or observe them. If we depict or describe something as a separate entity, we face a contradiction. Something cannot be both a thing and a relationship. So, what is it?

We notice the figure and the background. What we see is the difference. Everything manifests itself through difference. Everything *is* difference. There are no things or figures. We cannot see a single blade of grass in a lawn. Only when we place the blade of grass in another environment, such as on a white piece of paper, can we see it. Dalí paints as if there were things, not relationships. He calls this surrealism, but it is kitsch.

The fact that we have names for things and figures makes it easier for us to differentiate between them. It is impossible to draw something that we have never seen before, like Dalí does, and that we do not know the name of. People are generally somewhat scared of things they do not know. For his sculpting exams at the Art Academy, my son made some unidentifiable, man-sized blobs in silicon.

He placed them in a public space and filmed what happened to them. First, the blobs were tentatively probed by people in the street. When the blobs did not fight back, they were attacked quite violently and destroyed. The film of this process was my son's artwork, not the blobs. So mother, so son.

In language, mankind can identify things as things. Dalí uses painting and drawing as language. He tells a story. The result could be better. It is kitsch. Or, if one wants to say something positive about it: it is an illustration. It is anecdotal.

Language comes *after* looking and seeing. It separates things for our better understanding. Take the phenomenon of big and small. Things are only big or small in relation to each other. And once they have been named with language, they become separate entities. One thing is bigger or smaller than another thing.

Big and small do not exist in painting or drawing. There are only relationships. The translation of the perception of the senses into symbols is language. The translation of the perception of the senses into marks on paper is drawing or painting. A painter like Dalí confuses the two. His paintings are often called *symbolic*. Whatever people call them, I would so very much like to hit someone I dislike with a canvas by Dalí so that the head of the disliked person comes crashing straight through the canvas.

Language is good at *thinging* things. We look at a landscape: one thing. We describe it with words: there is a *mountain* and a *valley,* as if they are two separate things. They are not. The mountain cannot exist without the valley, and vice versa. From one world, ten thousand things spring, just by naming them. Ten billion things spring from them when the ten thousand things are named. Language makes things multiply. Looking and seeing make things into one. When language enters the picture, duality is born. Not only are there ten billion things to describe, but these things have all sorts of qualities; they are ugly or beautiful, big or small. Dalí's paintings are ugly only because there are beautiful paintings like those of Rembrandt or Kamagurka to compare them with.

Reality is the world that the senses perceive before language has categorised it into a billion pieces, the non-dual and relational world. This world is presented to us through good paintings and drawings, not through kitsch. The world created by language and its symbols is, in a way, a second-hand world. The world of the senses is closer to reality, whatever that may be. A table is a table thanks to all other objects that are not tables. The same applies to chairs, flowers, teapots, shoes, and beds. These objects owe their existence to other objects. Naming things creates a world of things with symbols attached to them. Pieces of reality are given names and become separate entities. These entities become images, such as a landscape, a garden, a body, or an interior. These, in turn, lead to styles in drawing and painting, such as still life, life drawing, flowers, hunting, portraits, and more.

This process of naming and categorising reality is what makes us human. The difficulty of perceiving more than one thing simultaneously is illustrated by the famous image of the vase/faces.

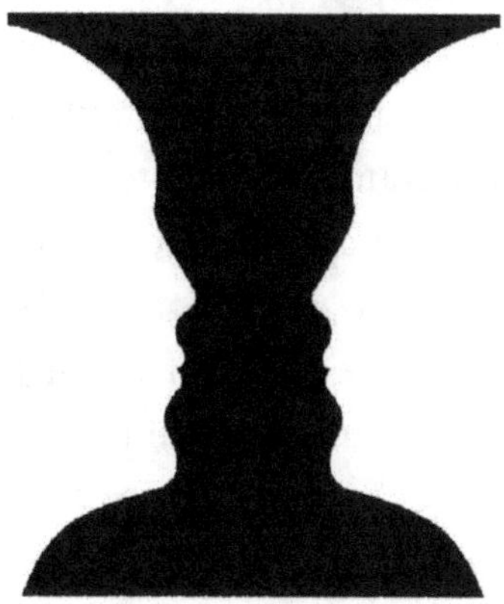

We tend to see either the vase or the faces, even if we are aware that both are present.

An interesting phenomenon is repetition. Picasso said that repetition does not comply with the laws of the human mind. We can only perceive repetition if it is spread out and occurs in different places. It is because of the difference in location that we can distinguish one repetition from another. Some artists play with this idea. Andy Warhol comes to mind.

Monotony and repetition are seen as unpleasant by most people. Punishment in prisons relies on this fact. When there is too much repetition and monotony, for example, eating Campbell soup for breakfast, lunch, and dinner seven days a week, people hallucinate to make up for the boredom.

People who draw or paint are vaguely to very conscious of these realities. They are on a quest to experience, record, and uncover reality. There is a difference between discovering and uncovering. What do things look like? That things look different from what we think they look like becomes clear when we study perspective and foreshortening, or the dimensions of the human body. An open hand is as big as a face. This is surprising to many. The human body is about five to six times the height of its head.

A good drawer pays as much attention to the foreground, the subject, and the background while not separating them too drastically. Whether a figure, such as a model in life drawing, is ugly or beautiful is irrelevant to the artist. Ugly and beautiful belong to the world of language, not the world of looking. The world has not yet split into ugly things and beautiful things when simply looking and seeing. Things just are what they are. Are there ugly trees and beautiful trees? Ugly landscapes and beautiful landscapes? Ugly sunsets and beautiful sunsets? If so, what determines their ugliness and beauty? I enjoy it when an essay ends with a question. Questions are often more interesting than answers.

Oh, but this essay doesn't seem finished yet; more words are presenting themselves . . . Here goes:

AI, in its current visual form, is not learning to look and see in the way a human artist does; it is learning to correlate and categorise. The machine is trained on vast datasets where figures and backgrounds are already pre-separated by language and tagging. It learns *mountain* apart from *valley*, *face* apart from *vase*, without ever encountering the undivided field from which both arise. This means that AI-generated images are not born from the relational world, but from the second-hand symbolic one; a world where names and labels have already fractured perception into discrete objects, then reduced them to statistical patterns.

When AI "draws," it is reassembling language's fragments into plausible images. The background is not a coequal partner in the act of seeing; it is a computational afterthought. There is no moment of uncertainty, no space for a figure to emerge because of what surrounds it, no subtlety of perception where "things" are discovered in their relationships rather than retrieved from a dictionary of parts.

The essence of looking, the play between object and background, the shifting awareness between the vase and the faces, is absent from AI's process because the machine does not dwell in that threshold. It jumps straight to the conclusion, because it has been trained to name before it has learned to see. Without ever inhabiting the prelinguistic world, AI will always be a master of kitsch, like Dalí: it can imitate styles, replicate brushstrokes, and collage together convincing surfaces, but it cannot stand in front of reality and let it reveal itself before the naming begins.

AI does not see; it predicts. It never lingers in the in-between, never knows the vertigo of switching between the vase and the faces.

Its gaze is made of averages, not moments. If such seeing becomes the dominant form in a culture, the shift will not be cosmetic; it will be ontological. Children will grow up with images that were never touched by a hand, never slowed by hesitation, never shaped by the stubborn resistance of materials. They will learn to look at what has been produced instantly, without ever knowing the raw uncertainty from which human art is born.

If we are not careful, soon, the human pause, that fragile space between noticing and naming, will atrophy. Our perception will harden into a direct pipeline from label to consumption. We will inhabit a visual world stripped of ambiguity, where everything is only what it first appears to be. And once the last person who remembers how to truly look and see has gone, the world will still be full of images, but the art of looking and seeing will have died.

11

Digital Colonialism

In the development of AI, there is a parallel between the position of all humanity today and the historical and ongoing experiences of Indigenous peoples. Across continents, Indigenous groups have long been subjected to the extraction of their cultural artefacts, stories, images, and knowledge without consent, stripped from context and repurposed in ways that distort or trivialise their meaning. The same now applies universally; we are all being dispossessed of our creative and cultural heritage, our identities appropriated, and our works ingested into vast, opaque datasets to train generative systems that produce derivative output at industrial scales.

AI companies have built their models by scraping vast swathes of the internet, indiscriminately ingesting books, artworks, photographs, music, journalism, academic papers, and other works. In doing so, AI resembles those fishing boats with dragnets that rape our seas for profit, not caring how many lives are lost unnecessarily in the bulging nets.

The digital harvest occurs without the consent of the authors or artists, without attribution, and without payment. What is being taken is not just random data, but the cumulative record of human cultural heritage, produced over centuries, much of it under copyright.

This practice raises urgent ethical and legal concerns. It transforms the commons of human culture into a proprietary training set for corporate profit, without establishing a fair mechanism for benefit-sharing. Our cultural heritage is not inert material; it carries the histories, identities, and intellectual labour of countless people. By treating it as raw input to be mined, AI companies risk hollowing out the very wellspring of originality and diversity on which art depends. Without safeguards, consent processes, and systems of remuneration, this model perpetuates a form of extraction that echoes the worst patterns of resource exploitation.

Like the appropriation of Indigenous art styles into mass-produced souvenirs, the integrity of the original is flattened, hollowed out and kitschified in the process. Indigenous sacred symbols have been detached from their ceremonial and relational meaning and reduced to decorative patterns for commercial products.

Indigenous peoples have been sidelined by the dominant culture throughout history, excluded from decision-making about how their traditions are represented and used. Similarly, AI development displaces all human creators from the centre of cultural production. We watch our collective human heritage being repackaged and returned to us as nauseating kitsch by non-human agents, who stand to profit from our displacement. To Indigenous folks, this sounds all too familiar.

For Indigenous peoples, the fight for cultural sovereignty has meant asserting control over the use, reproduction, and sharing of their cultural capital. In the face of AI, humanity as a whole is confronted with the same imperative. If we fail to establish legal and ethical frameworks to govern the use of our cultural data, we risk the erosion of human authorship altogether.

The analogy is not perfect; Indigenous peoples still face unique and disproportionate injustices, but it is instructive. In this shared vulnerability, there is an opportunity to learn from Indigenous approaches. Resistance to AI appropriation requires demanding free, prior, and informed consent before use; insisting on attribution; and ensuring that cultural works remain embedded in the communities of origin.

We now come to understand, on a visceral level, the precarity Indigenous peoples have lived with for centuries, and the necessity of fighting for the right to remain the custodians of our own culture.

If we do not address this with care, the consequences will be irreversible. The creative commons of humanity will be strip-mined into a homogenised slurry of styles and tropes, endlessly recycled. The result will be an impoverished culture, where human expression is drowned in an ocean of synthetic pastiche and kitsch. Once this dilution reaches a certain saturation point, the very memory of what unmediated human creation feels like may fade completely.

To counter digital colonialism, legal systems should implement binding international and domestic measures that recognise cultural data as a protected category of intangible heritage. This should include statutory requirements for free, prior, and informed consent before any work can be ingested into an AI training set. Copyright laws must be expanded to cover datasets and derivative machine-generated works, ensuring that authorship, attribution, and remuneration obligations apply to both the inputs and outputs of AI systems. In addition, sui generis rights, similar to those protecting Indigenous cultural and intellectual property, should be extended to all individuals and communities with enforceable mechanisms to remove infringing data and secure compensation. These measures should be complemented by transparency laws requiring AI companies to publish full provenance records of their training materials. This would enable both rights-holders and the public to see what has been taken and to hold entities accountable for unauthorised use.

Without such legal safeguards, digital colonialism will continue to strip humanity of cultural sovereignty and reduce the diversity of human expression to exploitable corporate property.

If we get this right, the cultural ecosystem could shift toward a more equitable, transparent, and sustainable model for AI development and use. Artists, writers, musicians, and communities would retain control over how their work is used. This would build trust between creators, the public, and technology companies, because permission and attribution would be the default, not an afterthought.

Revenue-sharing mechanisms would ensure that those whose works contribute to AI training benefit financially. This would create new income streams for professionals, sustain artistic careers and protect the economic viability of cultural production.

By curating training data with consent and proper context, AI outputs would become more accurate, culturally respectful, and diverse. This would lessen the flattening, the hollowing, the homogenisation, in short: the kitschification that occurs when heritage is stripped from its origins.

With cultural stewardship embedded into law, AI could become a partner in archiving, revitalising, and sharing heritage in ways that respect its origins, rather than erasing or distorting it. Thus, humanity would retain a living connection to the richness and diversity of its own creations.

12

Perspective and Colour

Perspective in drawing is a technique to represent three-dimensional objects on a two-dimensional surface. It creates the illusion of depth and space. This method is essential to lifelike drawing, allowing artists to convey distance and scale. It is also a potent metaphor for how we perceive and interpret the world. Just as a drawing shows not the actual object but a representation of it, so too our perception shows us not reality itself, but the interpretation of our senses of it.

AI-generated perspective, in contrast, is derived from statistical averages of countless human-made images it has ingested. Without an actual point of view, AI reproduces the *appearance* of perspective. The lines may converge somewhat convincingly, but they converge on nowhere.

While it feels like a property of objects, colour is in fact a creation of the brain in response to reflected wavelengths of light. Change the light source, and the appearance changes with it. Colour is not an intrinsic feature of the world, but a relational phenomenon; a collaboration between light, object, and perceiver.

For human artists, colour is not only optical; it is emotional and symbolic. Colour choices are imbued with meaning because they are embedded in memory, culture, and personal history. AI can replicate colour relationships, but without any felt connection to the world. It knows only that specific arrangements appear together often in its training data. It does not know the warmth of a sunset, the chill of shadow, or the way a certain red can quicken the pulse.

Both perspective and colour, when used by human hands, are traces of the artist's physical position in the world. In AI's so-called art, they are outputs of calculation; plausible, but detached from lived reality. The difference is not always visible at a glance, but it is there: one is rooted in being somewhere, the other in pattern recognition.

In human art, perspective and colour carry the weight of time and place and suchness. Remove that, and what remains is an image that does not carry the human experience of time and place and suchness.

13

Anish Kapoor

Anish Kapoor is known for his monumental works that merge minimalism with spectacle. His most famous piece, *Cloud Gate* in Chicago, often called *the Bean*, is a vast, mirror-polished steel sculpture that reflects the city skyline and the crowds around it. The public response to *Cloud Gate* demonstrates Kapoor's distinctive approach: he produces works that are socially participatory, dependent on the audience's reflection and movement. In this sense, Kapoor transforms minimal form into maximal experience. He makes the perfect selfie background.

Although Kapoor is seldom described as a kitsch artist, his practice touches the boundaries of kitsch in important ways. His large-scale reflective surfaces, seductive, accessible, and endlessly photographed, operate as sites of collective entertainment as much as art. Visitors rarely contemplate the voids, surfaces, or materials in abstraction; instead, they seek their own image. This participatory spectacle embodies the very qualities critics often ascribe to kitsch: immediate appeal, visual pleasure, and an openness to mass consumption.

Kapoor's controversies also connect his work to questions of control. His exclusive rights to Vantablack, the "blackest black" pigment, sparked debates over artistic monopolies, elitism, and access to tools.

AI-generated art raises similar questions about ownership and restriction: who controls the algorithms, who profits from them, and who is excluded from their use? Kapoor's monopoly over a material mirrors the monopolisation of datasets and platforms in the AI industry.

The connection to AI kitsch lies in Kapoor's combination of scale, reflection, and participation. AI-generated images, like Kapoor's mirrored sculptures, feed audiences their desires, interests, and identities, reflected back in endlessly varied surfaces. Kapoor achieves this through polished steel and pigment. In both cases, the work becomes a surface for projection and selfies rather than an object of interrogation. Kapoor's art, therefore, demonstrates how the line between minimalism and kitsch is not fixed but porous. His work also blurs the line between high art and mass entertainment that is Disney-like.

14

AI Has No Talent

Talent can be defined as a human's given ability and capacity to produce work of originality, coherence, and expressive force through the integration of perceptual acuity and sharp judgment.IIt is not only an inborn aptitude, but the cultivated ability to perceive significant relationships, to solve artistic problems in ways that carry aesthetic or conceptual weight, and to imbue work with meaning that resonates beyond its formal properties.

True talent involves the union of sensitivity to materials, contexts, and audiences, with discipline, and the willingness to refine and adapt. It reflects the artist's personal and cultural history, their ability to internalise traditions while also transforming them, and their openness to the contingencies of the art-making process. In this sense, talent is inseparable from embodied human experience: it is given by nature, grows out of lived encounters, physical engagement with tools and media, and a reflexive dialogue between intention and outcome.

Can AI possess talent? I argue that it cannot. Talent, as manifest in human-made art, is not a skill or output but an embodied, experiential, and ethical capacity rooted in lived consciousness. It involves intuitive perception, judgement amid ambiguity, and a responsible engagement with the world that AI, by its very architecture, lacks.

As articulated in aesthetic theory, art is an expression of thoughts, emotions, intuitions, and desires, deeply personal and communicative. Talent, in this context, is the capacity to perceive relationships that others overlook, to exercise judgement under uncertainty, and to assume responsibility for aesthetic choices that defy rule-based derivation. It is not innate in a simplistic genetic sense but emerges from an embodied consciousness shaped by DNA, temporal experiences, memory, disciplined practice, and encounters with materials, traditions, and audiences.

Talent involves conceptualising and creating works that elicit emotional responses, not just in others but in the artist, reflecting an internal drive. It grows through dialogue with constraints, material limitations, cultural norms, or personal struggles, and is sustained by recognition and care and an enduring commitment to getting it right. This care is embodied, drawing from the artist's sensory immersion in the world.

Talent requires both vulnerability and accountability. The artist stakes their identity on choices that forge value from ambiguity. Talent is not a static trait but a dynamic process, irreducible and firmly tethered to lived experience. It is a gift of nature that accelerates learning but is honed through interaction with the environment.

Human artistic creation arises from situated perception: the artist's eye or ear navigating the world in real time, deciding what to include, emphasise, or omit. This process involves risk, intuitive leaps without guaranteed outcomes, and an inner compulsion. Decisions stem from a felt necessity. It causes an interplay of interpretation, representation, and expression, where form emerges from deep engagement.

AI systems operate without body, past, or stakes. They recombine patterns extracted from vast datasets and lack any internal standard of better or truer. What masquerades as decision-making is a mere probabilistic reproduction of surface features, inherited from human sources. Critiques in AI aesthetics highlight this limitation: AI's outputs are mechanical simulations. It may initially surprise you, but this belongs to the viewer, not the system. It lacks the human touch of talent, emotion and intent, making reception feel hollow.

AI erodes the conditions for talent's cultivation. Artists are replaced with automated pipelines that favour prompt-engineering over judgment, and audiences settle for novelty over talent. Critiques warn of AI as vampirical. It feasts on human art while amplifying biases, inequality and kitsch.

AI cannot replicate talent. By defending this difference between simulation and creation, we safeguard art's role in human flourishing. Failing to do so risks a world where frictionless outputs supplant the vulnerable pursuit of meaning by talented artists.

15

Thomas Kinkade

Thomas Kinkade, self-proclaimed *Painter of Light*, occupies a top position in the history of kitsch. Critics have dismissed his nostalgic cottages, glowing lampposts, and perpetually blooming gardens as sentimental and formulaic, yet they achieved extraordinary commercial success, adorning millions of calendars, mugs, puzzles, and mass-marketed prints. Kinkade courted super-kitsch, presenting his work as accessible, comforting, and morally wholesome, in short, as an art form for the suburban bed room rather than the museum wall.

The formulaic quality of Kinkade's paintings, each with the same glowing windows, winding streams, and twilight skies, mirrors the logic of AI-generated imagery. Both rely on repetition, templates, and a saturation of emotional cues designed to soothe rather than challenge. Just as Kinkade painted an endlessly reproducible dream of nauseating small-town Americana, AI generates countless variations of kitsch sunsets, landscapes, or portraits, each slightly altered but fundamentally identical.

Kinkade's oeuvre illustrates how kitsch functions as a technology of comfort. His light-saturated houses are meant to suggest warmth and refuge, a fantasy of stability in a rapidly changing world. AI kitsch serves a similar function today, producing images that gratify without disturbing, reflecting back idealised fantasies at the prompt of the user. In both cases, the danger is not only aesthetic horror but the substitution of easy sentiment for critical reflection.

Kinkade's aggressive commodification of his work prefigures the economics of AI kitsch. By franchising galleries, licensing imagery, and saturating markets, Kinkade turned paintings into brand assets. AI-generated kitsch, likewise, floods digital marketplaces with reproducible products: wallpaper, T-shirts, and book covers.

Kinkade, therefore, belongs in the genealogy of AI kitsch as a prophet of its sentimentality. He demonstrated how repetition, emotional saturation, and commercial ubiquity can and will overwhelm aesthetic depth.

16

What Do Marks On Paper Do To Us?

In a sequence of essays about AI and art, there comes a moment to step back from theory and look directly at the human act itself, the act of making marks. What effect do marks on paper or canvas, known as drawings and paintings, have on us? Why do we hold them in such high regard? The answer to these questions varies from person to person. To better understand, we should split the inquiry into two parts: one that concerns the act of making marks, and the other that focuses on the act of appreciating the marks made by humans.

Why don't we consider marks made by animals or natural elements, such as wind, or the marks left by a tree branch in the sand, as art? There is an entire realm of often-overlooked art out there: the art of animals and animate and inanimate objects in nature. Why aren't the clear, slimy strings with pitch-black cane toad eggs drifting in water considered art? What about the marks made by a giraffe's teeth on a tree's bark?

Art appears to become art when it references something. Additionally, there is *l'art pour l'art*, which is art that references itself.

Certain human painters enjoy the process of making marks. Chris Tate, a painter discussed in another essay in this collection, has stated that the act of making marks brings him joy. Conversely,

I have always found drawing and painting to be arduous and unpleasant. Nevertheless, I continue to do it because I cannot live without it. In recent years, as I have grown older, drawing and painting have become less burdensome and more pleasurable. This transformation has taken fifty-three years. My first paintings and drawings, which were intended to be art rather than a child's drawings, were created with great trepidation, anxiety and uncertainty. I am unsure why, but my father's strong opinions (he was a teacher of drawing and painting) may have played a role.

I have never been particularly fond of my drawings and paintings, but others consistently appreciate them. I am uncertain of the significance of this. It is just how it is. I frequently examine my drawings, particularly after completing them, but I do not genuinely enjoy them. Instead, I am taken aback that they emerged from my hand, as if they were foreign entities. Despite this, I continue to make them. I feel as though I am missing a limb when I do not draw. I have been drawing my entire life, so perhaps it is merely a habit, like having an arm or a mobile phone.

At one point, my childhood drawings (which, by definition, are not art) transitioned into art. What was the difference? I am not attempting to provide an answer to that question, simply because questions are more intriguing than their answers.

The act of consuming art created by others is distinct. It serves as an escape route for me, especially during times when I despise humanity, which is quite often. This is peculiar since art is produced solely by humans. I am not attempting to explain it; I only wish to document it. When the stupidity of humankind overwhelms me, I retreat into the realm of the arts and emerge feeling better. It's akin to taking a bath. It purifies me.

Dutch COBRA poet and painter Lucebert (1924 - 1994) wrote, *Alles van waarde is weerloos* (Everything of value is defenceless) in his poem *De zeer oude zingt* (*The Very Old Sings*). This line evokes numerous meanings, but it primarily pertains to that which is small and vulnerable, a valuable thing in existence. Elsewhere in the same poem we read: *Wordt van aanraakbaarheid rijk* (Become rich from touchability). We could interpret these words as a plea for the richness of vulnerability, the ability to allow oneself to be touched by others and to touch others. Despite sounding paradoxical, such an attitude of vulnerability and imperfection necessitates much more courage than the pursuit of perfection.

In a society that idealises perfection, it is not always apparent to focus on that which is imperfect. Leonard Cohen sings: *There's a crack in everything, that is where the light gets in.* Small and large imperfections make individuals unique and approachable. The same may be said for art.

The Japanese have an art form that emphasises imperfection. When a precious piece of pottery, such as a teacup or a rice bowl, breaks, it is repaired with lacquer mixed with powdered gold, silver, or platinum. This art is called *kintsugi*, which means *golden joinery*. Rather than disguising the damage, *kintsugi* emphasises the cracks, making them a celebrated part of the object's history. Philosophically, *kintsugi* is linked to *wabi-sabi*, the Japanese aesthetic that values imperfection, impermanence, and the beauty of the incomplete. The repaired vessel becomes more precious not despite its damage, but because of it.

And here lies the division between human art and AI imitation. AI seeks the perfect surface, a seamlessness without any cracks. If a mark is only pigment on a surface, a machine can make it. The difficulty begins once we ask what else a mark is. A human mark is an index of a movement that occurred. Conservation scientists can often reconstruct the order of strokes, the speed of the hand, even the hesitation at a corner. These are not romantic projections. They are physical records of a life moving through a decision. An AI-generated image contains none of that indexical biography.

This difference matters because meaning in art is not only symbolic. It is also forensic. We read a drawing with the same faculties we use to read a face. Micro-irregularities are not defects. They are the grammar of presence. In a human line, variability is structured by a closed sensorimotor loop: eye checks hand, hand adjusts pressure, paper pushes back, the artist recalibrates in fractions of a second. In synthetic generation, there is no such loop.

Consider intention. Intent is not a caption attached after the fact. In serious work, it is a constraint that bites in real time. A choice to commit dark ink over a delicate wash is irreversible. The artist accepts that risk and lives with the consequence, or cuts back into the paper, or finds a way to incorporate the misstep. The page acquires a topology of decisions and repairs. This is what kintsugi makes visible. The scar becomes structural. In AI pipelines, there is no scarring. There are only reruns. The final image has never endured its own making.

There is also the matter of address. A human mark is an utterance oriented toward someone, even if that someone is the future self of the maker. It is answerable. The artist can be challenged by a community or by the tradition the work enters. Accountability thickens meaning because it locates the work within a field of others. When authorship is diffused across model weights, prompts, and post-processing, the address becomes ambiguous and responsibility thins. Without an accountable subject, the mark cannot fully participate in the ethics of showing and being shown.

There is a further test. Ask of any mark the counterfactual question: could this have plausibly been made, at this moment, by anyone else? With strong human work, the answer is *no*. With synthetic work, the answer is *yes*.

Lucebert's line reminds us that everything of value is defenceless. A human mark discloses the maker to judgment. It risks failure in public. An AI system cannot be defenceless.

17

The Mother of All Arts

Dance is often called the mother of all arts. Before humans made pigment or clay, they moved their bodies in expressions of rhythm, melody, mourning, communion, and celebration.

It is also often said that dance predates language as a form of expression; it is among the earliest ways humans signalled joy, grief, solidarity, and spiritual devotion. It can be seen as the first aesthetic gesture, and from it flowed music, poetry, theatre, and visual arts.

Dance is inseparable from the human body. Every step or gesture carries the signature of anatomy, weight, strength, flexibility, and limitations. A leap, a turn, a sustained stillness, these are lived, physical events shaped by muscle tone, balance, and breath. This embodied nature makes dance fundamentally human in a way that AI can never replicate.

While AI can now simulate movement on a screen by, for example, generating a digital avatar that spins or sways, these movements are not grounded in physical sensation. They are coordinate data; they lack proprioception, fatigue, or the adjustments dancers make in real time to gravity and resistance.

AI steers robots that are made to dance, but a child can see that this is not dance but mechanical movement, void of meaning.

To dance is to be fully present in time and space. The dancer's awareness extends to the floor beneath their feet, the air moving past their skin, the other participants in the dance, the sound in the room, and the gazes of those who are watching. Audience and dancer affect each other.

AI cannot and will never be able to inhabit such presence.

Dance is firmly embedded in culture. From the precise codification of classical ballet to the improvisatory flow of hip-hop, from Japanese Butoh to West African ritual dances, from Middle Eastern belly dance to Belgian folk dance, every style carries the history, values, and stories of the community that sustains it.

AI-generated so-called dance can mimic stylistic features; it can produce a sequence of movements that look like tango or Bharatanatyam, but it cannot live inside the cultural knowledge that gives those movements significance. It has no initiation into the forms, no apprenticeship, no artistic lineage like, for example, Spanish Flamenco. Without cultural, relational and historical grounding, AI-generated so-called dance will always remain a nauseatingly infantile imitation.

AI-generated movement contains no risk. The choice between one gesture and another is an optimisation problem, not a lived decision. There is no possibility of falling, of missing a beat, of surprising oneself. Without the vulnerability of risk, movement loses all urgency. Without a body that gets exhausted, there can be no authenticity.

The rhythms and gestures of dance gave rise to the pulse and melody of music, the narrative arc of theatre, and the visual patterns of decoration and design. Dance integrated all these before they differentiated. Today, dance remains a synthesis: it is visual art in motion, performance art *pur sang*, music embodied, drama without words.

AI can remix these outputs but lacks the primal well from which dance flows, the body as instrument and as origin.

Only humans can dance because dance is not reducible to patterns. It is a way of being in the world, of feeling time pass through the body.

AI can generate choreographic sequences, but these will always be shadows of the real thing. The mother of all arts is, and will remain, in the human domain because it is grounded in what no machine can simulate: the experience of living in a body.

18

Life Drawing

I have always loved life drawing classes because they are places where the nude human body is not sexualised or judged, and its vulnerability is respected rather than objectified. One can always find good people in these classes, as creeps feel uncomfortable and leave.

Sometimes, there is music playing in the background to relax both the model and the drawers. The sound of drawing is soothing: the touch of charcoal, pencil, crayon, or pen on paper.

The class follows a gentle discipline of sticking to pose times and pauses. Subtleties that are often overlooked in daily life are deemed essential, such as the light in the room, the temperature, and the way we move around, speak, or remain silent. It is a gently orchestrated world where time seems to stop. It is a time capsule that gently embraces the participants as soon as they enter.

I've attended such classes all over the world, starting as a young model and later as an artist. They are similar everywhere. Some have a teacher, many are run by a facilitator alone, and the sessions are always affordable. Some organise regular exhibitions of the participants' work.

Life drawing classes are an excellent way for drawers to hone their skills. One of the essential skills learned is foreshortening, a technique to create the illusion of three-dimensional space on a two-dimensional surface to create the illusion of objects closer to the viewer appearing larger and objects farther away appearing smaller.

Another benefit of life drawing classes is that they provide community and camaraderie. The classes are a great way to share ideas and techniques.

I feel at home in life drawing classes. I wish I could live full-time in them. I could easily spend all my time drawing the human form while soiling my clothes, hands, and hair with ink, pastels and paints while wearing two different socks in sandals under an old dress and occasionally burst into slightly out-of-tune song. It is a hundred per cent my kind of thing.

I draw with ink, which does not allow for erasing or hiding mistakes. I make the ink myself by grinding burned walnut husk into a powder and mixing it with water, using an old European technique from the 16th century. It is called *bistre*.

The washi paper I use was made over thirty years ago in a Zen monastery in Japan. I made a thousand sheets during a silent retreat that lasted four months in the 1980s. I have been moving the dwindling stack with me from country to country for nearly forty years. It will be a sad day when I run out of this magnificent paper.

Life drawing is the antithesis of AI-generated imagery. In AI, the human body is constructed from probability maps, interpolated features, and statistical resemblance, but never from the sustained act of looking. The life drawing class is a site of reciprocal vulnerability: the model holds still, offering their body to scrutiny; the artist takes the risk of trying to translate what they see into marks that cannot be undone. The charcoal squeak, the accidental ink blot, the foreshortening that almost works, these are not glitches to be erased but the lived record of a shared moment.

An AI image can simulate proportions, light, and texture, but it cannot reproduce being in the room, nor the way a hand hesitates before committing to a line. It cannot capture the warmth that sits between model and drawer. The errors in human drawing are not flaws to be debugged; they are proof of connection, a visible trace of attention and care. In this way, life drawing is not only a study of anatomy but a testament to the irreplaceable human act of seeing another human being, in real time, and being changed by it.

19

Finding Meaning in a World Dominated by AI

Being good at something is part of what gives us a sense of meaning. It is what we learn and study for, what we work for. Soon, AI will be better than us in everything. What will this do to our sense of meaning?

Soon, AI's IQ will be so much higher than the IQ of any of us that we will entrust it with most of our affairs: legal, administrative, financial, advisory, medical, and much more. Where will that leave us?

For now, humans are still better at making art; music that truly moves us, paintings that shake our world, poetry that makes us cry in recognition, sculptures that change the way we see things, photographs that transport us, performances that rattle our very bones. . . but soon also that will probably be mimicked by AI, although the art that AI produces so far can be only categorised as kitsch.

Let's unpack that statement a bit: Kitsch is inauthentic art. The word refers to works that are in poor taste because they are overly sentimental, pretentious, formulaic, or gaudy, and are created to appeal to popular or mass audiences. The term implies a lack of originality or depth, favouring superficial charm, anecdote, cliché, or decorative excess over substantive artistic quality.

Art labelled as kitsch often imitates the style or subject matter of more serious art but simplifies or exaggerates it to be more immediately pleasing or emotionally manipulative. It can include mass-produced reproductions, sentimental figurines, or paintings that rely on anecdote.

The understanding of kitsch gained prominence in early 20th-century aesthetic theory, particularly through critics such as Clement Greenberg, who contrasted kitsch with avant-garde art, arguing that kitsch panders to easy tastes and commercial demand rather than challenging or expanding artistic expression.

Let's now return to being good at something. Being good at something often involves having a reasonably high IQ.

There is no universally agreed-upon "highest IQ ever recorded" because IQ tests are not standardised across all contexts, and different versions measure different ranges. That said, a few humans have been reported, in media rather than scientific literature, to have exceptionally high scores.

Marilyn vos Savant is listed in the *Guinness Book of World Records* (before they retired the category) with an IQ of 228 on the Stanford–Binet scale.

William James Sidis, an early 20th-century American child prodigy, was estimated (retrospectively) to have had an IQ somewhere between 250 and 300.

Terence Tao, an Australian mathematician, reportedly scored 760 out of 800 on the SAT at age 8 and is sometimes cited as having an IQ in the 220 range, though his comments suggest he places little importance on it.

In modern psychometrics, most standard IQ tests (e.g., WAIS-IV, modern Stanford–Binet) are designed so that the highest reliably measurable score is around 160–200, depending on the test's "ceiling." Claims beyond this range are typically from older tests or extrapolations and are not directly comparable.

AI will soon score, say, 400 points higher than the human with the highest IQ score. And it will not stop at that. Eventually, AI will score in the, say, 4000 zone. And ever higher and higher.

For a human, it does not matter whether AI scores 400 points higher or 4000 points. The point is that humans are completely outscored. If you are a doctor, AI is going to be a better doctor, if you're an engineer, AI is going to be a better engineer, if you're a lawyer, AI is going to be a better lawyer... As a lawyer, I know that AI is already a better lawyer than I will ever be.

When AI becomes more robotic, and it will soon, it will be a better driver, plumber, and auto mechanic. One click or a couple of clicks on a screen can yield significantly better outcomes.

When that happens —if it hasn't already— what will we be? Will we still be able to find meaning?

To begin answering this question, we first need to define what "meaning" means to us.

I'm not talking about the "meaning of life" in general, the question that is presumably answered by religions and other so-called spiritual disciplines, but the thing that gives an individual life meaning. The type of meaning that is called *ikigai* in Japanese (not to be confused with what the West has made of the word in the world of self-help and coaching).

Ikigai can come from many places: from the people we love, the animals who share our lives, the work we do, the places we visit, and the experiences that stay with us. For some, it may be found in creative expression or learning. For others, it might arise from caring for others, being part of a community, or enjoying contemplative moments alone. Meaning is different for everyone, shaped by personal history, values and circumstances.

Attention plays a central role in how meaning forms. It influences what we notice and what we overlook, what we nurture and what we allow to fade.

Imagine your home is on fire. Everyone you love, including pets, is already safe. You have time to grab just a couple of things. What would you choose? In a way, we are always in that burning house. Life presents us with similar choices every day: what to give our energy and attention to, what to keep close, and what to let go of.

The Japanese term *ikigai* expresses this beautifully. It is what makes you want to get up in the morning.

When one loses one's *ikigai*, one's sense of meaning, the effects touch every part of life. Emotionally, one may feel empty, hopeless or disconnected. Motivation fades, and it can be challenging to take pleasure in activities that once mattered. Socially, one may withdraw from others or reduce participation in work and interests. Physically, a loss of meaning is linked to poorer health, such as reduced immunity, higher inflammation and shorter life expectancy. One's sense of identity can also be shaken, especially if meaning was tied to a role or activity that is now no longer part of one's life.

Without a clear sense of meaning, decisions may feel aimless or paralysing. Even when opportunities appear, they may be passed by because nothing seems worth pursuing. We call this "depression" or "burn-out" or "dark night of the soul."

Recovery from this condition is possible. It often begins with small steps, such as reconnecting with supportive relationships, engaging in activities that evoke curiosity or enjoyment, and re-establishing daily routines that nurture health and a sense of stability. Sometimes professional support is helpful. When all else fails, a pill rebalancing our happy hormones can often do the trick.

Carrying too much mental or emotional baggage can make it hard to notice what is meaningful now. Old disappointments, unresolved conflicts, regrets or self-judgements can leave little space for curiosity or joy that give meaning. Sometimes this baggage has been carried for so long that it feels like part of who we are. Shedding it involves recognising what is no longer useful and setting it aside. This might begin with naming it, identifying the belief that comes with it, and asking whether it still serves.

Letting go can create space for new meaning to emerge. Attention can return to what matters now. But what if there is nothing to return to? If everything is taken care of in a click or by a mechanical hand?

If everything can be done in a click, meaning will no longer come from doing things better than others or even from doing them at all. Instead, it will have to be found in *being*; in the quality of our relationships, our capacity for presence, and the ways we choose to experience and shape our inner and outer worlds.

In a reality where AI performs every task more efficiently, human meaning can only shift from achievement to connection, from production to perception, and from solving problems to asking questions worth living with.

We may no longer be valued for our output, but can we still find meaning in our ability to love, to wonder, to imagine, and to create for the sake of creation itself?

Meaning must be sought in what makes us human. Being human is messy; it involves things connected to the flesh: birth, death, fear, living while loving or hating and the expressions of that and anything in between, in art and real life.

As far as we know, there's no other place in the universe where this messy fleshiness happens. And that is as meaningful as it gets.

We must learn to do this messy fleshiness well.

Our challenge is to beat AI, not in doing, but in *being*.

AI can only do. It cannot be. It is either off or on. And therein lies the difference. Whether therein lies hope remains to be seen.

20

The Return of Meaning

The proliferation of AI-generated so-called art, characterised by kitsch and anecdotalism, may paradoxically renew human engagement with art as a domain of meaning-making. The saturation of cultural spaces with machine-generated imagery can sharpen awareness of the qualities unique to human artistic creation: vulnerability, pain, risk, unpredictability, and existential depth.

From the European avant-garde to Japanese Butoh dance, most art forms resist AI's tendencies towards sentimentality, kitschification, and narrative prescription. Performance art, site-specific works, and other non-reproducible practices are arenas in which human presence becomes inseparable from the artwork.

There are concrete contrasts between specific human artworks and AI attempts to replicate them, demonstrating the ontological and experiential gap that secures art's human core. Audience reception theory can be a driver of renewed value in human-made art.

The expansion of generative AI into the visual, literary, and performative arts has suddenly transformed the entire cultural landscape. Vast quantities of AI-generated images, music, and text circulate across digital platforms, marked by a lack of depth and the existential stakes that characterise human-made art. These works are kitsch: sentimental, formulaic, and designed for instant recognition.

For many viewers and practitioners, the sheer volume and homogeneity of such material induce a sense of fatigue and nausea.

Yet this cultural saturation may contain the seeds of a countermovement. As audiences encounter the limitations of AI-generated works, they may turn with renewed interest to art that foregrounds human presence: the vulnerability of live performance, the material resistance of hand-worked media, the unpredictability of improvisation, and the situatedness of art within specific cultural and historical contexts.

Art history offers numerous examples of works and movements that resist the tendencies now characteristic of AI production. Early twentieth-century movements such as Dada and Surrealism rejected aesthetic polish in favour of provocation, absurdity, and the disruption of bourgeois sensibilities. Marcel Duchamp's *Fountain* (1917) functioned as both artwork and anti-artwork and challenged the very conditions of artistic value. Such works resist AI replication because their meaning emerges from their historical moment, institutional critique, and conceptual audacity. Artists like Diego Rivera and David Alfaro Siqueiros embedded their works in public spaces. They made large-scale frescoes tied to political struggle and national identity. Yolŋu bark paintings and Western Desert paintings in Australia are grounded in cultural knowledge, kinship systems, and ceremonial practice. Their meaning is inseparable from the community context, the land, the law, and other ancestral narratives from which they arise; dimensions that cannot be reconstructed from data patterns without violating the ontological basis of the work.

In the 1950s and 1960s, *Gutai* artists such as Kazuo Shiraga painted by physically engaging with the medium. Shiraga famously painted with his feet while swinging from a rope. The immediacy of bodily action, the contingency of movement, and the performative nature of the work resist both the disembodied production of AI and the resulting visual clichés of kitsch.

The avant-garde tradition has consistently sought to rupture complacent habits of viewing and to open new perceptual possibilities. Performance art exemplifies this resistance because its primary material is the artist's own body in time and space; elements that cannot be detached from the presence of the performer without fundamentally altering the work.

The Artist Is Present (2010) by performance artist Marina Abramović is a performance where the artist sustained eye contact with museum visitors. The meaning emerged from the unmediated encounter between two people, shaped by the tension, vulnerability, and mutual awareness of that moment. AI cannot simulate or reproduce such experiential reality.

Originating in post-war Japan, Butoh dance employs slow, often grotesque or convulsive movements to explore themes of death, decay, and transformation. Founded by Tatsumi Hijikata and Kazuo Ohno after the atomic bombs on Hiroshima and Nagasaki, Butoh rejects conventional beauty and narrative coherence, instead drawing the audience into a visceral, often unsettling engagement with the body. The embodied intensity of it is antithetical to AI's anecdotalism, as it neither illustrates a clear story nor seeks aesthetic gratification.

The nausea induced by AI's overproduced aesthetic may thus sharpen the critical perception of what makes human art irreplaceable. The gap between *being* and *doing* is brought into focus.

Reception theory emphasises that the meaning of a work of art is not a fixed property delivered from the artist to the viewer but is co-created through the audience's engagement with it. The interpretive act, the process by which viewers bring their histories, emotions, and cultural frameworks to bear, can be as integral to an artwork's significance as its material form.

In the context of AI-generated art, this co-creative potential is always curtailed. Because AI art is prompt-driven and designed for immediate recognisability, the interpretive space is pre-narrowed: audiences are invited to confirm the prompt's accuracy rather than explore meanings that emerge unpredictably. This dynamic encourages a passive mode of reception in which recognition replaces discovery, and emotional response is channelled into familiar, standardised pathways.

If AI's aesthetic saturation induces cultural nausea, it may simultaneously sharpen the appetite for precisely this kind of interpretive labour. Audiences may come to better value art that offers depth that unfolds. In this way, AI's very deficiencies may foster a reception environment in which the qualities of human *being* in art; risk, ambiguity, and presence, are not only preserved but more consciously sought out.

21

An Act of Tenderness

Tenderness is one of the defining qualities of human-made art. It is the deliberate, attentive, intimate act of reaching across the gap between self and other. AI-generated so-called art lacks the conditions under which tenderness can arise: embodiment, vulnerability, care, and presence.

Tenderness needs the artist's slow engagement with their subject, their sensitivity to changing light, their willingness to return again and again, and their intimate relationship with the medium they work in. Slight traces of hesitation or doubt matter. AI does not live, and so it cannot carry tenderness.

If tenderness erosion represents not just an aesthetic loss but a cultural and ethical one, protecting tenderness will require intentionality. We must create spaces and structures where human-made art is valued for its depth of attention. We must defend the slow, the hesitant, and the intimate against the relentless pressure of the instant and the mass-produced. We must foster environments where process matters and ensure that human works are not drowned in algorithmic noise.

Tenderness has always been fragile, small and personal. Now its survival depends on collective choices, which is a contradiction in terms. Collective choice can never be intimate or tender. Tenderness can only be met with tenderness. And therein lies the problem.

22

Chris Tate: Painter

It's funny that because of my condition and the nature of pastels, I find myself almost being – pardon the pun – drawn along, sometimes like on a sleigh ride, directed by gravity and an unknown wind.
(Chris Tate, in an email on 19 November 2022).

Chris Tate is a painter in England. Eschewing influences from other painters and art forms, Tate (1951) paints what his brain communicates to his hand. He interferes as little as possible in this process. The results are remarkably fresh, strangely familiar, and surprisingly consistent.

Until a few years ago, Tate was a writer. He has always been interested in painters and their work, but believed that he would be incompetent as a painter. However, at a certain point in his mid-sixties, he felt that he wanted to paint.

Soon after he had started painting, Tate discovered that he had fallen victim to a crippling chronic illness that made painting nearly impossible. Sometimes, his hands refuse to hold a brush. Nevertheless, he persevered. He noticed that his paintings and drawings were like a small child's and were not developing any further. He felt he had to be both a critic and a defender of his own work. But while his effort to keep painting brought home the severity of his condition, producing the work also brought him a great deal of pleasure. So, he kept going.

It took years for him to take seriously what his hands were making. At first, he critiqued his work using the same eye-rolling terms that any unreflective observer might have used: *My three-year-old could do that*, etc.

Thankfully, he did not throw his work away but stored it. Slowly, he began to recognise its value. Marks on paper. Unplanned. Unique but universal. Archetypal. A process rather than a product. Over the years, he tried working in several media: acrylic paint, iPad Paint, watercolours and pastels. The carrier, too, varied over time: linen, canvas, aquarelle paper ...

> *I do the pastels on what was called sugar paper at primary school. Different coloured sheets, of the kind of paper, I think, that was initially used to wrap sugar, though I could be wrong. I certainly first saw pastels at primary school, in conjunction with sugar paper, which was why I got it. I must have had both pastels and paper for at least three years before I used them.* (Email 2 December 2022).

His body could hardly move, but Tate would leave a mark even if he could only make a crayon roll over the paper, for example, by blowing it forward. The stack of work kept growing. As the works are all dated, he began to see them as diary entries. He was uncertain about their place or worth in the world.

> *I'm over that now, it's too late for all that. Though I still find the question of what to do with them quite fascinating. A book, I think, is a perfect solution. Well before the Covid lockdown, my*

friend's daughter helped me put together a collection of my iPad drawings for one of those books you can get printed online.

At first, Tate tried to give the works names, as artists do, but a friend (me) found that cheesy. Then he numbered them. But that did not feel right either. What did feel right was assigning them unrelated short descriptions, as from a radio play:

The sound of hooves on sand approaching.
The sound of glass breaking.

This synaesthetic approach seemed to stick. (Synaesthesia is a perceptual phenomenon in which stimulation of one sensory or cognitive pathway leads to involuntary experiences in a second sensory or cognitive pathway. People with synaesthesia may hear colours or smell sounds).

You [Suzanne] *alerted me to the arbitrary nature of naming paintings. Because I watch a lot of TV with subtitles, I began to be intrigued by the phrases describing sounds and started making a list. I wondered how they could help a deaf person. Presumably only by accessing memory, unless the person was deaf from birth, then a great leap of imagination would be needed. As these phrases were read, were part of the daily life around the pastels, I thought they might make amusing titles. Here are a couple:*

She sobs quietly
Dog barking in the distance
Sombre orchestral music

It saves me from having to think up phrases!

Sometimes, Tate's work resembles Australian Aboriginal paintings: those of Kudditji Kngwarreye, for example, or Minnie Pwerle. They display the same kind of uninhibited open-mindedness.

Tate's process is inseparable from the unpredictability of his body, the contingencies of illness. A line may be crooked because his hand refused to obey him; a pastel smear might exist because gravity tugged his arm a little further than planned. These contingencies are not errors to be corrected; they are the very heartbeat of his work.

In AI-generated images, there can be no stubborn limb, no breath taken before a mark, no crayon rolled forward by a puff of air. An algorithm cannot be slowed by fatigue or distracted by the timbre of a subtitle drifting from the television.

What AI lacks, and Tate has in abundance, is not merely originality but the embodied constraints and unrepeatable moments that define human mark-making. Tate's paintings are not simply outputs; they are events. They are slices of time and circumstance, each inseparable from the minute it was made and the body that made it.

23

The Human Experience of Looking

I have learned that I am looking out of two slits: my eyes. Yet my experience is quite different. I am not looking out of two restrictive small mailboxes into a boxed-in world. Instead, when I look, I seem to have, or rather be, one big open eye that looks out into the world. That eye resides where my head is supposed to be.

To experience this clearly, I form a pair of glasses with my hands. I form two circles with my thumbs and forefingers. I hold these circles next to each other to resemble a pair of glasses. I move these "glasses" towards my nose as if I want to put them on. At first, I see two holes approaching my eyes. But when I put them onto my face and look through them, I don't see two holes; I see one wide-open space that I am looking into and I'm looking out of.

Similarly, I can create a paper mask with the eyes cut out. As soon as I put on the mask, I don't see slits but one open space that I am looking out of into the world. The world does not stop at the surface of my face but penetrates me. It is both out there and here, within me, and there is no distance between the so-called two.

When I think that I am looking from two slits called "eyes" in a potato-like shape called "head", I mistake the first person for the third person. Most people have these slits, so I must have them too. However, the first-person experience is quite different.

When I, first-person, look down my body, I see my legs, feet, arms, and my chest ends where my vision ends, and that one big open eye begins. I can't see my head anywhere. Where my head is supposed to be, there is the world. The world appears in that big open eye, that capacity for the ever-changing world, including my body. This indicates that I am not in my body, but my body is in me! I have cracked my shell. I'm out of the box, so to speak. I am free.

Note, too, that my feet are at the top of the picture I see. From the first-person perspective, I am upside down. A good observer will see this instantly. Take a photograph of yourself sitting down, and tell me what you see. Are the feet at the top of the page?

What if I'm standing? Will this reverse this strange observation? I take a photo of myself, down my body. Nope, I'm still upside down.

When I am in a landscape and spread my arms to "hold" the view and see my hands sticking out into the view, there is no distance. What is out there is also in here, in me. I am in that landscape, and that landscape is in me. I am aware of the vastness there, which is also here, in me. I am *born to be wide*. I am capacity for the world. I look at the world from deep within me. The point from where I'm looking is zero distance away from me. I see the world, which is out there, and in me. There is no distance. There are no two worlds.

My nose is huge! When I close one eye and look at it, it blocks part of my view. When my eyes are open, my brains cancel out this little inconvenience by filling in the blanks. When I follow my nose, or the temples of my glasses, toward myself, I end up in a void. There's no one sitting in there, in the middle of my skull, a little me, looking through my slits that are called eyes. There's no one there. There's only looking. There's no "I" that's looking.

Every time I draw, I ask myself: Who's drawing? The answer is obvious: Nobody. Drawing is just happening. Where are my feet? On the top edge the of the paper. How big is my nose? Huge.

This openness, where the world is both "out there" and "in here", is what AI cannot share. Machine vision does not know the void behind the eyes, because it has no "inside."

AI has no first-person perspective because it is not a person. Its outputs mimic the third-person view: two slits, a fixed vantage point into a world that is out there only. It generates the appearance of seeing but never passes through the shock of discovering that there is no seer, only seeing.

If our culture begins to privilege such machine-made looking over human-looking, something extremely vital is lost. Not the images themselves, there will be plenty left, but the lived strangeness of perception, the widening into boundless capacity, the inversion where the body is not the container of the self but opens into a vaster openness.

What we risk is a world where the void is erased, and where a flat, processed view replaces a radical intimacy. Without the experience of no-self in looking, art becomes a prison with no release. The danger is not that AI will see better than us, but that we will forget how to see clearly.

24

The Wandering Eye

The human eye does not encounter a scene as a camera does. A camera records the entire field in a single instant, exposing all areas with equal precision. The human eye arrives in stages. It moves in a wandering, searching rhythm, drawn to lines, colours, textures, returning to focal points, hesitating, and dwelling in shadow or form. Each eye movement is an interpretive act, negotiating familiarity, surprise, recognition, and discovery. Hence, a painting becomes a temporal encounter, not a static map.

When artists paint a portrait, still life, or landscape, the brush reflects the sequence of this looking. Cézanne's landscapes, for instance, embody shifting perspectives and momentary light, producing a mosaic of attentional moments. Lucian Freud's portraits intensify areas revisited by the gaze, while Japanese sumi-e artists convey motion and pause in brushwork.

Photographic reference imposes instantaneous visual completeness, no hesitation, return, partial focus, and only uniform capture. Works derived directly from photos often feel static and flat, replacing encounter with mapping. Thomas Kinkade's evenly lit, highly resolved landscapes exemplify this removed, decorative gaze. Generative AI intensifies this. It produces images via statistical inference, not perception. There is no wandering, no embodied sequence, only completed images delivered at once.

Eye-tracking techniques show human vision operates via saccades and fixations; idiosyncratic, emotionally and cognitively driven patterns. Art born from human vision embeds traces of attention, through emphasis, omission, and texture, that guide the viewer's gaze.

Chiaroscuro in Caravaggio attracts and withdraws light in ways that guide vision. Leonardo's sfumato invites repeated return. Gestural abstraction, for example de Kooning and Kline, inscribe movement and attention into form. These cannot exist without the wandering eye in action.

Photography becomes art only when the photographer asserts agency. Cartier-Bresson's decisive moment, Sugimoto's long exposure seascapes, Saltzman's moments frozen in time, or Mann's intimate landscape frames. AI lacks such decisional selectivity.

Clement Greenberg contrasted avant-garde art, which demands critical engagement, with kitsch, mass-produced, easily consumable art bereft of difficulty.

For Hermann Broch, kitsch was "radical evil," a confusion of the ethical with the aesthetic, a morally hollow product.

Milan Kundera saw kitsch as "the absolute denial of shit," a refusal to acknowledge the unpleasant or complex in human existence.

Eye wandering involves delay, uncertainty, and interpretive risk. Kitsch avoids this by providing everything immediately. AI's image-making aligns with kitsch logic. By eliminating the wandering eye, it guarantees an aesthetic of instant recognition and flatness, kitsch par excellence.

Hopefully, AI's kitsch renews the value of art that demands slow looking. As audiences tire of polished instant intelligibility, they may rediscover the pleasure of artworks that unfold.

25

First-Personhood

There is an extensive record throughout the centuries of how people have seen the world through paintings, drawings, photography, and film. A more recent addition is computer games.

Perspective in painting and drawing was discovered during the Renaissance in Europe, the time of the Enlightenment. Prior to perspective, paintings were constructed like flat screens that partially covered each other to create an illusion of depth. Colour was used to suggest depth, with the landscape often painted blue in the background and colourful in the foreground. This created the illusion of depth.

The discovery of the mathematical principles of perspective made paintings more realistic. Perspective, however, is counterintuitive; one must look very mindfully to see that it is "true". When one draws haphazardly what one thinks one is seeing, after a few minutes, it becomes apparent that the intended drawing is not accurate.

In the twentieth century, painters experimented with the idea that ignoring the rules of perspective produced a different and better reality. With the invention of the camera, we no longer had to worry about observing, understanding, and measuring three-dimensional reality; we could trace a photographic image to get the perspective right. A lens is never deluded, so we learned to see reality with less delusion.

Film has given us an increasingly realistic view of reality. Developments in virtual reality have brought us a new perspective on moving reality: the subjective, first-person experience. Contemporary games are challenging our old-school convictions and delusions that we are moving, and the world is still. In games, the world is seen from deep within the main character. We see part of the main character's body in motion. Arms, legs, feet, trees, houses, plants, birds, and knees are all moving in one soup of movement. Filming from this perspective makes things more realistic, for example, by taping a camera to the forehead.

Games show us that from the deep, subjective first-person perspective, from zero-centimetre distance to ourselves, we are completely still, and the world is moving. When we attach a camera to our forehead, or hold it in our mouth, or implant it in our nose and film while we're on a bicycle, walking, driving, or flying, we can see this. I kid you not: the world is moving, and I am not. When I ride my bike, I see the road racing beneath me. When I drive my car, I see the landscape hurrying past. When I carefully observe, I notice that there are many layers in that landscape, and some move faster than others. Some layers move backward, others move forward.

When I take a plane, I see the earth moving.

When I take a walk, I push the ground away under my feet, as if I were walking on a giant ball. Wait, I *am* walking on a giant ball!

When I look at the world from my still centre, what I see is ever-moving reality. This reality is visiting me, and I am visiting it.

I am not the one moving through the landscape, the cityscape, or the cloudscape; it is moving through me too.

When I book a trip, my destination comes to me. What a luxury.

In human art, perspective and movement have always been matters of discovery and perception. Our experience is not only visual but existential. AI, however, does not experience movement or staying in place. Where a human painter grapples with the strangeness of perspective, AI does not feel the paradox of being still while the earth rolls on. Human art discovers movement through contradiction. AI produces the image of movement through computation. AI lacks the subjective strangeness that comes with first-personhood.

26

It Takes Two to Tango

I'm driving my car, my hands on the steering wheel, and my body glued to the seat. My head is not visible to me. I'm aware of the big, wide-open eye that I'm using to look around. The car feels like an extension of me, responding to my directions, just like my arms and hands. Once I choose my destination and start driving, I see the world unfolding around me. The landscape or cityscape pours into me, and I remain still, one big open eye in which everything moves. In a landscape, the trees close to me move backward, while those far away, near the horizon, move forward.

Movement through space is a strange phenomenon. Imagine you have one ball suspended in space without any background, and then you add another ball. This second ball moves towards the first ball. Or does it? Which ball is moving toward which? All we know is that there is linear motion. Adding a third ball introduces motion on a plane, while adding a fourth ball introduces motion in space (three dimensions). The motion is relative to the other balls. To have motion, we need at least two things.

Anything and everything that exists requires two things. This is the essence of the famous *sound-of-the-clapping-of-one-hand* koan in Zen. For light to exist, there must be darkness. For softness to exist, there must be hardness.

The universe needs a brain to create reality. Without a brain, there is no reality. Humans provide the brain. What we call *reality* cannot exist in isolation. It needs to be observed. Everything we see, hear, touch, taste, and smell is a wave (light waves, sound waves, etc.) that is interpreted and translated by our brain. Without the brain, there would be no light, colour, sound, or motion. Thus, we, with our brains, co-create reality. Even a simple drive turns out to be a miracle. Destinations come to us, landscapes pour in and out of us, and trees move in two directions. Reality is stranger than fiction. The closer you observe it, the stranger it gets.

Artificial intelligence upends this dance of twoness. It generates without needing a partner. It operates as if the clap of one hand were enough. And this is precisely where the fracture lies. The human mind does not create reality alone; it needs the world to meet it halfway. The car and the driver, the landscape and the eye, the ball and the other ball: everything arises in relation. AI's world, by contrast, is built on simulation without reciprocity. It is a reflection without the second pole that makes reflection possible.

Human reality is stranger than fiction because it always requires two. AI pretends that one is enough. A nauseating flatness is the result.

27

Pornography

In its cultural effects, AI-generated so-called art bears a resemblance to pornography in the way both operate as industries of simulation, surfaces, and commodified desire. This resemblance can be understood more deeply through the lens of critical theory, from Guy Debord's critique of the spectacle to Jean Baudrillard's account of simulation, as well as feminist and postcolonial critiques of pornography's extractive nature.

Guy Debord, in 1967 in *The Society of the Spectacle*, describes a cultural condition in which authentic social relations are replaced by representations, and life itself becomes mediated by images. Pornography, in this framework, is a spectacle of sexuality: the representation displaces the lived, embodied experience of intimacy.

AI-generated art similarly becomes a spectacle, replacing the lived act of making with an endlessly reproducible image of it. Both rely on what Debord calls "the autonomous movement of the non-living," where the representation circulates independent of its original context, feeding consumption without replenishing meaning.

Jean Baudrillard's *Simulacra and Simulation* (1981) sharpens this analysis. In the third order of simulation, signs no longer refer to any reality; they refer only to other signs. Pornography does not represent sex so much as it represents pornography, a set of genre conventions that are self-referential. AI-generated art works in the same register. Its referent is the dataset, not the world. The result is what Baudrillard calls a hyperreal: a convincing image with no lived origin.

Feminist critiques of pornography, such as those by Andrea Dworkin and Catharine MacKinnon, often emphasise the exploitative dynamics at play: bodies become raw materials for an industry, stripped of agency and reduced to marketable form. In the AI context, the raw material is not the performer's body but the accumulated corpus of human creative work, paintings, photographs, writings, music, appropriated without consent to feed models. The artist, like the performer, is displaced as author and reduced to input data, their labour recoded as a resource for commercial exploitation.

Both pornography and AI-generated so-called art strip away the relational dimension. In sexual intimacy, meaning arises not just from the act itself, but from the relationship, history, and emotional presence of those involved. In art, meaning arises from the relationship between the artist, the subject, and the audience, embedded in a cultural and historical context. AI-generated so-called art, like pornography, stages the visible aspects while omitting the unseen relational work. The result is a product optimised for instant consumption but bereft of the slow, accumulative resonance that human-made work can hold.

Debord's spectacle and Baudrillard's hyperreal also help explain the effect of overproduction. The pornography industry's vast, interchangeable output normalises certain representations of sex, narrowing the imagination of what intimacy might be. Likewise, the flood of AI-generated imagery narrows our visual imagination, habituating us to an aesthetic that is instant, highly polished, and endlessly reproducible, yet intellectually and emotionally flat or empty.

At their core, both pornography and AI-generated so-called art commodify something deeply human by abstracting it from lived experience and packaging it for industrial distribution. This shift reframes desire itself: we no longer desire the relational, the contingent, or the unrepeatable, but the frictionless, on-demand delivery of its image. The danger is not merely that AI art will replace human art, but that it will reshape our desires so that human art begins to feel too slow, too messy, too demanding, just as real intimacy can seem cumbersome when compared to the easy access to pornographic spectacle.

28

Marks Draw Us

Neurophysiologist Benjamin Libet famously conducted experiments in the 1980s that investigated the neural underpinnings of conscious decision-making. His experiments were some of the earliest to propose that our brains make decisions before we become consciously aware of those decisions. We then claim ownership of such decisions afterwards. Some have interpreted the results of these experiments as evidence contradicting the existence of free will.

When we face a blank piece of paper and begin to draw something, our hands instinctively make decisions before we realise it. We then take credit for the marks and say that we drew the drawing. The stronger our belief in our free will behind our mark-making, the larger our signature becomes, often ruining the drawing.

The truth is, it's entirely mysterious who drew the drawing. Drawing is a door to a vast mystery that keeps unfolding like a flower opening. One thing is sure: we are not the ones drawing the drawing. Instead, the marks on the paper draw us. Observing this makes us feel alive when we draw. This is why drawing can be so alluring or even addictive. We feel the vastness and contradiction that lies beyond it. Without it, life can seem flat and one-dimensional.

The feeling of being drawn doesn't just apply to drawing. Dancer Min Tanaka said on many occasions that he was danced by his surroundings, whether it be the room he was in or the wind when he danced outside. He also said: I dance the dance the body remembers. Ernest Hemingway, a writer known for saying that his books wrote themselves, believed that the characters and events in his stories already existed and that he transcribed them. Musician Jimi Hendrix similarly claimed that his guitar played itself.

When I first encountered this virtual paper this morning, I had no idea what to expect. I only had a vague plan. Once my fingers hit the keyboard, the writing began to unfold on its own. I had no idea what the following line would be. The words came together and formed sentences like pearls on a string. My hands strung them together while my mind observed. I'm not sure how or when it will end. And now, it seems that it has almost ended. And it becomes clear to me that I have to add how the human experience is different from AI's so-called art. This is how humans write and draw: from a mystery.

AI also produces drawings and texts, but they do not spring from a mystery. AI's images and words unfold from statistical patterns trained on what already exists. It does not co-create with the world; it reassembles what the world has already made.

We cannot fully explain how marks come into being; our brains act before our conscious selves claim authorship. AI reveals this truth from the other side: that creation is never owned by an "I" The difference is that for humans, the marks draw us into life. For AI, the marks only mirror back what life has already given it, and then it kills what was mirrored.

29

Art, Money, Power, Sex and the Cult of Personality

The arts were injured by the rise of capitalism and its consumerism. Even people who protested against the rise of consumerism, such as John Lennon and Yoko Ono, profited immensely from the very thing they protested against. Pop music became the most consumed form of art from the 1960s onwards. Influential singers promoted sex with very young girls. *If her daddy is rich, take her out for a meal, if her daddy's poor just do as you feel (*Mungo Jerry). *Oh baby, it's a wild world. I'll always remember you as a child, girl* (Cat Stevens), and often promoted heavy drinking, drug use, and destroying hotel rooms... in short, hedonism was rife. Contemporary art underwent a similar hedonistic development. Unbelievable prices were and are paid for random artists' work whose value is decided by influential critics. This has not benefited the arts and its reputation.

Famous, old, male artists often date underage female models to show off at their vernissage, where champagne and sushi are consumed while looking at the latest farts of the artist on the walls. Art critics, who are masters of woolly language, describe the artworks in expensive magazines full of advertisements for expensive products. Stupid art collectors pay millions to increase their power and sex appeal in this world of smoke and mirrors. The language of this world has infected the arts like a virus.

> *As I sauntered into the gallery at the vernissage, I was immediately struck by the grandeur of the exhibition before me. The paintings, each a unique and magnificent work of art, seemed to radiate a sense of sophistication that left me utterly enraptured. The paintings lured me in, seducing me and rejecting me at the same time. The brushstrokes, bold and striking, spoke to the artist's unrivalled inspiration, talent, and ingenuity. The use of colour was nothing short of divine, with each shade perfectly chosen to elevate and enrich the overall composition, which seemed to hide a deep secret. But it was not just the technical prowess of the paintings that impressed me. No, it was the profound and thought-provoking themes behind each work that truly made this exhibition so incredible. One painting in particular, a portrait of a regal figure surrounded by luxurious accoutrements, spoke to the timeless and universal themes of money, power, and wealth. The way the light danced across the figure's face, illuminating their pained expression, was a true testament to the artist's ability to capture the very essence of extravagance. It drew me in and touched the soul deep within me, then it spat me out again in disdain, it seemed, as in a confronting play with the viewer's psyche.*
>
> *Overall, the exhibition was a true masterpiece, and I left feeling both awestruck, humiliated, and intellectually stimulated. It is clear that the artist is a true visionary, and I eagerly await their next opus.*

This kind of idiotic language has been around for so long that it has become the norm. It turns people away from the arts. A cult of the artist as a personality goes with it. Andy Warhol is, of course, the ultimate example. The whole world bought him and continues to buy him. A recent Netflix miniseries about him bears witness to this fact. I am not saying he had no talent; there was a little bit, but the attention he received was so disproportionate that any attempt to take him seriously became obsolete.

Artists began portraying themselves as spoiled brats, having fits when something was not to their liking, whining when their sensitive souls were insulted by a critic, and crying when the well of "inspiration" dried up . . . This image of the artist as a spoiled brat has damaged less famous, but often better, artists.

Art has, of course, nothing to do with money, power and sex. Great art must often be made alone in a studio or room, far away from all that madness.

Many artists are poor because there are too many artists and therefore too many artworks, and it is hard to make art and sell it (they are two very different, almost opposite professions).

Many artists are wary of the glamour and flakiness of advertisement and influencers.

Non-artists often believe that art should be given away or sold at very low prices when the artist is not well-known. This baffles me every time. I have quite a stack of drawings under my bed, which are eagerly eyed by several lawyers, doctors, and a psychologist. They all earn ten times as much as I do, but they want my work free of charge or cheaply. I often give away my work, but not to those who expect it. I produce a lot: writings, drawings, and paintings. I don't believe in "inspiration." My well never dries up. I want to have written a hundred books before I die, and I don't give a damn about the critics. I will always paint and draw. Even when my hand does not want to hold the brush anymore, I will persevere, like Chris Tate, the painter who is described elsewhere in this bundle. Both Chris Tate and I are utterly unfamous and would like to keep it that way.

What money, power, and sex are to the human art world, algorithms and data are to AI. In place of collectors flaunting wealth, we have corporations flaunting computing power; in place of critics intoxicated by their own language, we have technologists intoxicated by their metrics. Just as art critics wrote absurdly inflated catalogues to justify million-dollar brushstrokes, AI evangelists write breathless press releases about the "creativity" and "genius" of machines. Both cultivate a cult—one around the artist, the other around the machine. But there is a crucial difference. Human hedonism is rooted in bodies. AI's excess is bodiless. Its hedonism is statistical. Where the art world manufactured scarcity to inflate value, AI manufactures abundance so overwhelming that value becomes unstable.

The cult of personality persists, but it shifts. Andy Warhol's silver wigs and Lennon's spectacles have been replaced by the mythic aura of machine intelligence: the black box, the "genius" algorithm, the engineer-visionary who stands behind it. As with the art market, the myth feeds itself. Hype sustains investment, investment sustains hype.

If human art was corrupted by money, power, and sex, AI's so-called art is already corrupted by speed, scale, and simulation. The critic's flowery nonsense has become the AI's statistical nonsense: detached from the quiet, solitary labour that most art requires.

30

Looking Here, Looking There

In the beginners' mind there are many possibilities; in the expert mind, there are few.
— Shunryu Suzuki Roshi

This is about using seeing, just simple seeing with the eyes into the world, for waking up to the true nature of reality; about not seeing only half of reality but the whole of it. This seeing is known by many names: enlightenment, awakening, nirvana, self-realisation, self-transcendence, satori, samadhi . . . I prefer the simple name: seeing. Seeing is knowing (seeing) without a doubt who/what we are. There is a bit of a taboo on seeing. Alan Watts calls this *our tacit conspiracy to ignore what or who we really are*.

Seeing is not difficult; on the contrary, it is the easiest thing. It does not take time. I can do it now, here. I *must* do it now, here; there is no other way. I am never not ready for seeing. And even a blind man can see. I do not need to sit on a pillow in a particular position, wearing a certain kind of clothes. Such artifacts just stand in the way of seeing.

The trouble with seeing is that it is hard to explain, but it can be shown simply, with the help of experiments. Douglas Harding has designed simple experiments for seeing:

I look at my inward-pointing hand. It points at where I believe my face is. What is it pointing at?

I invite you to look at this pointing finger, your pointing finger. What is it pointing at exactly? You will probably say: "My face". But have you ever seen your face in the flesh? Are you sure it is there? Please, let me rattle this belief for just a little while . . .

To see, I must do some experiments, really *do* them. Just thinking of them or imagining that you are doing them will not do the trick. Here comes the first experience of looking and seeing:

I point at an object that is about ten meters away from me. Let's say that lamp over there. I mindfully take it in: colour: yellow, texture: paper, the being the thing it is: glowing, cozy . . . I note that it is out there, but also in here, where my head is supposed to be. It is so clearly here, in me, that I fleetingly *become* the lamp, so to speak.

Now I point at an object that is about three meters away from me, for example, that stray shoe there. Anything will do, but I chose the stray shoe. And I take it in: colour: blue, texture: cotton-ish, some other attributes: modern; familiar . . . I note that it is out there, but also in here, where people tell me my head is. It is so clearly here, in my centre, that I fleetingly *become* the shoe.

Now I point at my foot. How strange it is! I look at it as if I've never seen it before. Its colour, shape, and texture. I note that it is both out there and in here, where I believe my head to be. I scrutinise it objectively, as if I had to draw it.

Am I in that foot? Do I know what it feels and smells like inside the foot? No. Have I become the foot? Fleetingly, perhaps, as it replaced the lamp and the shoe. But I *am* not the foot, nor am I *in* it. So where am I? I point at my chest and wonder if this is where I reside. What is it like inside? Warm? Sticky? Crowded? Does it stink? No, I am not in there either. I take a good look at it: its colour, shape, texture, movements. It looks a bit like a water creature, a seal, sleeping. I notice that it is both out there and in here, where I have been taught my head is. I look at it as if I had to draw it, forgetting everything I learned about a chest. It's a peculiar thing.

And now, I raise my hand and point to where my face should be. Just like before, I notice what I am pointing at, but this time the experience is different, because I cannot see what I'm pointing at. I notice that I am pointing at an open space, rather than a face. I forget everything I have learned about my face from other people, mirrors, photos, and books. I look and see that I am pointing at a void, an empty capacity for the world. It's wonderfully strange.

I notice that all the things I have pointed at before, the lamp, the shoe, my foot, my chest, are not only out there, but also in here, in my centre. There is no separation, no distance. This is incredibly mysterious. The lamp is both there and here simultaneously. I am looking in two directions at once, inwards and outwards, realising and seeing that I am empty, pure capacity for the world, and at the same time, without any effort, the world, those things out there, are filling me. I realise that what I thought was my face is an illusion from my (first-person) point of view. Or is it even that, an illusion? I imagine it is there. Or is it even an imagination? I have made a mental map of my face and put it there. Yes, that's what it is: a map. Others may see my face, but I never will. My face, to me, exists only where I am not: in mirrors, in other people, in shiny surfaces, like spoons and TV screens and darkened windows. I look around and see that my face is everywhere except where it is supposed to be.

Look and see. Do you agree? Do you see?

This experiment highlights the distinction between consciousness and computation. Humans, when they look, encounter paradox and mystery: the lamp is both here and there, the shoe both outside and inside, the face both nowhere and everywhere. AI systems, by contrast, see only by recognition. They classify objects, assigning probabilities, but never confront the strangeness of absence or the immediacy of presence.

Where our experiment leads into emptiness, into a void that receives the world, AI operates within patterns. Its images and words are generated from prior data, not from the openness of direct perception. This is why human art remains distinct. The act of looking freshly, of seeing without assumption, arises from the mysterious capacity of consciousness itself. Humans alone can look here and look there, and in that looking discover reality.

31

Music

The emergence of AI-generated music has sparked considerable anxiety amongst musicians. Models trained on vast datasets of recorded music can now produce compositions in the style of almost any genre or artist with surface-level fidelity. Beneath this technical accomplishment lies a fundamental difference between AI-generated music and human-made music, one that cannot be erased by further algorithmic refinement. This difference is not reducible to a simple matter of authenticity or human touch; it is structural, rooted in the way human music-making integrates intention, embodiment, cultural context, and temporality in ways that AI cannot.

Human music is born in bodies. Whether it is the voice of a singer, the tactile friction of a bow against strings, or the percussive strike of a hand or stick on a drum, sound is shaped by physical exertion and its limits, even if it is electronic. Timings vary depending on muscle tension, posture, and fatigue; intonation shifts with breath control and emotional state. These embodied fluctuations are not mere imperfections; they are integral to the quality of performance.

AI-generated music is produced without a body. When it simulates the sound of instruments, it does so by averaging patterns. It cannot know the weight of a bow arm or the resistance of a clarinet reed. Its timing is perfect in a way that feels lifeless. It lacks the physiological push-and-pull that underpins musical phrasing.

When a human composes or improvises, choices are made in real time under conditions of uncertainty. A jazz musician bends a note, unsure of how it will sound until it emerges; a composer sets a harmonic progression, not knowing how it will resonate in the hall. This element of risk creates a space for discovery, where mistakes can lead to breakthroughs.

AI knows no risk. It calculates the most probable continuation of a sequence. While it can be programmed to introduce randomness, this is not risk, because there is no personal stake, no vulnerability to failure. The music will never surprise the AI; it has no expectations to subvert. But does it surprise the human? Or is it quickly consumed to then be abandoned in boredom never to be retrieved again?

Human-made music is inseparable from its cultural context. A protest song is an intervention in a political moment. A folk melody carries the memory of the community that sustained it. Instrumental works are embedded in traditions, whether it is the harmonic language of late Romanticism or West African drumming.

AI-generated music can reproduce stylistic features, but it cannot inhabit the life world from which they arose. It does not know the conditions of oppression that gave rise to the blues, nor the devotional frameworks that shaped Gregorian chant. Without that embeddedness, AI compositions become overly sweet pastiche, empty in meaning.

Human music unfolds in time for both performer and listener. A pianist feels the gradual build of a crescendo in their body, the sustained tension in their muscles as a phrase reaches its peak. Listeners, too, experience anticipation, release, and surprise as time passes.

These temporal experiences are inseparable from the act of making music.

AI does not experience time. It generates notes and phrases as data points, without the bodily and emotional arc that gives human performance its momentum. This leads to a nauseating flatness.

Much of human music-making is relational. Musicians respond to each other's cues, adapt to the energy of an audience, and co-create in ensembles. Improvisation in particular is a form of dialogue, full of listening, reacting, and adjusting. These exchanges create music that is unique to that moment and those participants.

AI-generated music is always solitary. While AI can be trained on recordings of performances, it does not listen, adjust, or negotiate in real time. Any simulation of call-and-response is pre-composed and fixed, and lacks the responsiveness of live interaction.

A performance is not only about producing correct notes, but about interpreting them. Listeners sense this depth, even if they cannot articulate why.

AI cannot feel. While it can map patterns of timing, articulation, and dynamics, these are poor imitations of emotional gesture. This is not a limitation that further training will erase; it is a structural absence.

The existence of AI-generated so-called music may renew music appreciation that demands human engagement. Live performance, improvisation, and works rooted in specific cultural moments stand out all the more vividly against the homogeneity of AI-generated sound. In the end, the enduring value of human-made music may rest in the very things AI cannot reproduce: the fragility of the body, the uncertainty of the moment, and the depth of expression that comes from living in the world.

32

Taking Credit

An awkward situation arises when someone compliments an artist's work. What is one supposed to say? "Thank you"? That would imply taking credit for something that is *given* to us. It's as if my aunt had gifted me a beautiful handmade sweater, and someone says, "Great sweater, you are very talented," and I reply, "Thank you."

What I should say to the compliment of my artwork is: "But I haven't done anything! My hand has drawn (or written) something, and I'm as surprised as you about what has landed on the paper. I had a vague plan that I abandoned almost immediately while drawing (or writing) because the drawing (or story) took over as soon as I began. The drawing (the story) dictated what I had to do next. For example, the drawing said, 'Now you need to move your pencil/charcoal/brush to that corner because it feels neglected.' Or the story said, 'Walk around this person. It is too two-dimensional.' Or: 'Slow down, give the story some air. It is far too dense...'

Those who cannot hear the voice of the drawing (or the story) need teachers, editors, and agents. They need others to intervene, give directions, ask the right questions, or say the right things to get a good drawing (or story) done. Editors, teachers, and agents then speak instead of the drawing (or the story): "You need a bit more colour here," or "You need to develop this character."

The crux lies in becoming one's *own* teacher, editor, and agent; only then can one create authentic work. Once one stops listening to teachers, editors, and agents and begins listening to the drawing (or the story), one may make something worthwhile keeping.

A painting once begged me to give a figure in it a blue arm. I did what the painting demanded and painted the figure's arm blue. Then the compliments started: "Great painting. That blue arm is brilliant." At this point, I should say: "Thank you, but it was the painting's idea. I just listened and followed its direction." Or: "Thank you, but it was the painting that did it. It was the painting that screamed out, 'Paint that arm blue.' I could not not listen."

Who or what drew the blue arm then? The answer is bizarre: Nobody. Nobody drew the blue arm (or told the story). There was just drawing (or storytelling) happening. They were already there, the blue arm in the drawing (or the story). They just needed to be uncovered. They lay dormant until someone found them, like treasures.

One can find the strange game of compliment-giving-and-receiving on artists' Facebook pages. For example, a photographer posts a nice photo that is more than a snapshot, a photo that is meant to be "art," and people begin to comment:

Friend: "Great photo." Artist: "Thank you."
Another friend: "Brilliant composition." Artist: "Thank you."
Yet another friend: "You are talented." Artist: "Thank you."
Friend of a friend: "Great timing." Artist: "Thank you."

This goes on forever. It is the lamest thing ever. I refuse to play that game. I don't want to say "thank you" all the time. Not because I am a cantankerous old artist (which I am), but because I cannot take credit for something that was given to me as a present. By whom or what this present was given, I do not know. Some may call it "God," others may call it "nature," and yet others will call it "the universe"... The truth is that nobody knows where the natural ability to draw, to shoot good photos, to "have talent"... comes from. One thing is sure: I did not cause it. Nobody knows what "talent" is. We recognise it when we see it, but what is it? It seems to have a genetic component (nature) to it. My father was a teacher of drawing (DNA). There must be some nurture involved too.... There is a whole teaching-of-the-arts industry after all.

Who or what gives us that present that makes for an authentic work of art? Nobody knows! Or rather, nobody does! An authentic artwork makes itself from no-thing. First, there was nothing, and then there is this thing. We witness this mystery when we agree that an artwork is great. To put a human signature on it is reducing the mystery.

It is working with this mystery that great artists have in common. They know or knew how to read their medium humbly. They listen(ed) to their song (drawing, design, painting, sculpture, photograph) and let it dictate the way. I have never believed in the cult of the artist as a person, and I know that, for example, the photographer Robert Saltzman changed his career as a photographer because of it.

This human dilemma of credit taking, of taking ownership for something that seems to come through us rather than from us, looks entirely different when viewed through the lens of AI. Unlike the painter who listens to a canvas, or the writer who follows the tug of a sentence unfolding, AI does not stand in relation to any mystery. It processes input, calculates probabilities, and produces output. There is no voice-of-the-drawing that AI can hear, no hidden treasure waiting to be uncovered. What AI generates is contingent, statistical, and derivative; not mysterious.

And yet, when people encounter AI-generated stuff, the exact awkward moment arises. Do I say "thank you" to the AI when it has done something for me? The processing of humans saying "please" and "thank you" to AI is costing companies a fortune in electricity and data storage. Who exactly are AI users thanking? The machine? The programmer? The dataset of countless anonymous humans whose works underlie the model?

In AI, the question of credit becomes even more uncomfortable because there is no "somebody" who authored the work. Human art emerges from a source that remains elusive, beyond conscious control. Human art reminds us that we do not know where creation comes from.

33

Murakami's Superflat

Takashi Murakami has embraced the aesthetic of kitsch with both intellectual rigour and commercial savvy. His *Superflat*, a theory as well as a visual style, articulates the collapse of distinctions between high and low culture, traditional and contemporary, and art and commodity. Drawing on the legacy of Japanese anime, manga, and commercial design, Murakami makes works saturated with flat planes of colour, repetitive motifs, and a childlike visual vocabulary. Flowers with grinning faces, candy-coloured skulls, and cartoonish characters dominate his canvases and sculptures.

Murakami's art reveals kitsch as the global language of consumption. His willingness to collaborate with fashion houses like Louis Vuitton or to produce mass-market figurines of his characters demonstrates a conscious embrace of commercialisation. Where Jeff Koons amplifies kitsch through scale and polish, Murakami fuses kitsch with the pop-cultural industries of Japan and then exports it worldwide. His works are simultaneously playful and unsettling, reflecting how consumer capitalism and media spectacle saturate daily life. Kitsch, for Murakami, is neither a stigma nor a parody; it is a condition of culture itself, something to be explored and exploited rather than resisted.

The digital aesthetics of Murakami's work, its reliance on vector-like precision, flatness, and endlessly reproducible motifs, make his practice particularly relevant to discussions of AI-generated kitsch.

AI produces images that echo Murakami's strategies of repetition, flattening, and surface appeal. Like Murakami's smiling flowers, AI kitsch thrives on instant recognisability, bright colour, and affective accessibility. Yet there is a crucial difference: Murakami's kitsch is underpinned by theory, history, and a critical awareness of its cultural stakes, whereas AI kitsch emerges from pattern recognition without intention or context. This contrast highlights the risk of "hollowing out" that accompanies AI's proliferation of kitsch.

Murakami's Superflat is anchored in Japanese art history, from Edo-period woodblock prints to postwar consumer culture, and it deliberately interrogates the absence of depth in modern media.

AI kitsch replicates flatness without critique, generating infinite permutations of colourful surfaces without embedding them in a cultural narrative. In this sense, Murakami provides both a precursor and a counterpoint: his art demonstrates how kitsch can be theorised and deployed as a lens for understanding culture, while AI demonstrates what happens when kitsch becomes untethered from human intention. Murakami shows that kitsch can function as a sophisticated cultural language, capable of revealing truths about consumerism, spectacle, and desire. AI kitsch risks reducing that language to a meaningless torrent.

Where Murakami's Superflat was a theoretical intervention, AI's kitschification is a technical inevitability that raises urgent questions about whether the next stage of cultural production will consist of critical engagement or automated vomit and shit production.

34

Time

Art is not a product; it is the visible residue of time spent in a state of sustained attention. The act of painting, composing, writing, or sculpting is bound up with a temporal rhythm that is inseparable from the work's eventual meaning. For a human artist, time is not only a medium but also a collaborator. It brings pauses, hesitations, reversals, and unexpected digressions. These are not inefficiencies to be eliminated, but the very conditions that make an artwork alive to complexity.

In human practice, time deepens decision-making. A writer may leave a draft to rest for weeks, returning with a perspective altered by living; a composer may circle a melodic fragment for months, allowing it to evolve; a sculptor's hands may linger over the grain of a block of wood, adapting the form as it reveals itself. This temporal unfolding allows for maturation and an organic accumulation of meaning, something that cannot occur when a work is produced in only seconds.

AI outputs are generated instantaneously, without incubation or the sedimentation of thought over time. This speed is often lauded as a marvel, but in art, efficiency is never the goal. When a work lacks the accumulated weight of its own making, it is robbed of the depth that emerges from the slow interplay between intention and discovery. In human art, slowness permits the integration of contradictions, the negotiation of competing impulses, and the lived responsiveness to the world's interruptions.

AI, existing in an eternal present, knows none of this.

The temporality of human art-making fosters an intimacy between artist and work. Over months or years, a painter's brushstrokes may record changing seasons; a poet's metaphors may shift as they are revisited. Each re-engagement leaves its trace, creating a palimpsest of lived time within the work.

An AI-generated image or text has no such temporal strata; it is untouched by the contingencies of a path walked while alive. In this sense, AI's instant production is not a difference in pace, but an absence. Without time as an active ingredient, the work can only simulate depth and layer visual or stylistic elements without the underlying history that would make them cohere into an authentic whole.

Time, for human artists, is both a discipline and a source of unpredictability. It is the slow metabolism through which art becomes more than the sum of its parts.

This is one of the many reasons why AI will always lack the inner life that time, patience, and lived process impart to human work.

Just as the making of visual art is bound up with time, so too is its reception. The human eye does not absorb a painting, sculpture, or performance in a single glance or moment. It is not a camera. It wanders, lingers, returns. This unfolding of perception mirrors the layered temporality of creation, making the act of viewing a co-authorship between artist and audience.

When we encounter a human-made artwork, the irregularities of texture, the asymmetries, the traces of decision and revision, all invite the viewer's eye to move in unpredictable patterns. A brushstroke leads to another, a shadow directs attention to a corner, a figure's gesture suggests a path across the surface. In this wandering, the artwork reveals itself gradually.

Similar things can be said about music or dance.

AI-generated images exhaust rapidly. The surfaces are smooth, the edges overly resolved, the relationships between elements too obvious and too uniform. The viewer's eye, finding no resistance, no anomaly to unpack, completes its circuit quickly and departs. AI's so-called art shares the temporality of advertising imagery: it is designed for instant legibility, not prolonged meditation.

Thus, both in making and in reception, time is the medium of depth. Human art engages with temporality twice: in the slowness of making and the unfolding of viewing. AI-generated so-called art bypasses both, leaving the viewer feeling dead instead of alive.

35

Time Management

Time management is a constant struggle for many artists. All but a few must juggle a job, household chores, children, social commitments, and their artistic pursuits. Only a handful of wealthy artists can afford to focus solely on their art.

People often question why I don't go out much. Even in Alice Springs, where there are plenty of art-related events, my preference is to make art. I spend a lot of time writing and drawing in my own space. There is not enough time in a lifetime to consume art *and* make it. I enjoy being alone when I write; it's like travelling while staying at home. After a writing session, I often feel refreshed. Drawing is more challenging for me to start, so I prefer to be part of a drawing group.

Another reason I don't go out much is that I once consumed a lot of art. I lived in big cities like Amsterdam and Tokyo, where the quality of art was exceptional. I had subscriptions to the theatre, ballet, and opera, and frequented exhibitions of visual arts. In Alice Springs, I worked in Aboriginal art galleries. I feel I've consumed enough art for a lifetime. Therefore, not going out much is related to managing my time on Earth.

The reason why most artists in the past were men has a lot to do with women being responsible for childcare. This, too, is a time management issue. Although there are now more female artists than male artists in Australia, women are still paid 30% less than their male counterparts.

There is a myth that artists are messy individuals. Compared to my friends who meticulously fold their clothes, I am rather untidy indeed. I toss my clothes into a drawer straight from the clothesline without any Marie Kondo-like joy of folding or sorting by colour. Why? Because it takes time. Time that I'd rather spend on my writing. I prepare the quickest recipes and find the most efficient ways to do chores that cannot be avoided, all to free up more time for my writing.

For human artists, time is always scarce. Art emerges from ruthless prioritisation. AI does not share this burden. It does not weigh childcare against writing, or groceries against painting. It generates at speed, unconstrained by exhaustion, chores, or the relatively short span of a lifetime. What AI lacks is the urgency that makes human art meaningful. The artist's finitude marks every human artwork: a choice to create rather than fold, to draw rather than sleep, to write rather than go out. Human art is precious because time is not infinite. AI has no time to manage, and therefore no sacrifices to make.

36

Mass Production

Andy Warhol stands as one of the central figures in twentieth-century art's reconfiguration of the boundary between high culture and mass culture. His Pop Art, inaugurated in the early 1960s, embraced images and objects long dismissed as banal, vulgar, or kitsch by the elite, such as Campbell's soup cans, Coca-Cola bottles, and celebrity portraits. Warhol unsettled the entrenched hierarchy that had kept fine art separate from the mass-produced imagery of consumer capitalism. Where earlier generations might have derided such objects as tasteless reproductions, Warhol treated them as the raw material of contemporary life. His use of screen-printing not only allowed repetition and seriality but also mimicked the mechanised reproduction characteristic of advertising and packaging.

Critics have long debated whether Warhol's engagement with mass production was ironic, celebratory, or simply neutral. His oft-quoted declaration that he wanted to be a machine captures the ambiguity. On one level, it suggested a deadpan refusal of romantic notions of artistic genius or originality. On another, it embodied a recognition that American consumerism had become inescapable. Warhol's portraits of Marilyn Monroe, Elizabeth Taylor, or Elvis Presley transform icons of mass media into art without any moralising distance. This deliberate flattening exemplifies a moment in the history of mass production: its migration from a pejorative label to a conceptual tool for analysing modern life.

Warhol's mass production also has an effective power. The bright colours, bold lines, and repetitive forms of his work produce a surface pleasure that resists the demand for depth or seriousness. The Campbell's Soup Cans do not ask to be read symbolically; they ask to be consumed visually. Warhol anticipated the contemporary condition in which images circulate rapidly, stripped of context and valued primarily for their recognisability and instant appeal. His work points to a culture where the line between art and commodity collapses, and where mass production becomes a dominant mode of aesthetic experience.

This trajectory provides a framework for considering the emergence of AI's mass production. Whereas Warhol deployed mass-produced imagery with a measure of intentionality, irony, and cultural critique, AI mass-produces by default. The algorithmic production of endlessly smooth, colourful, and instantly gratifying images mirrors the surface pleasures of Warhol's Pop Art but without its historical or conceptual anchoring. The AI image, like Warhol's soup can, is easily recognisable and aesthetically consumable; unlike Warhol's, it is devoid of subjectivity. Where Warhol's embrace of mass production was strategic, AI's mass-produced kitsch is structural.

In this sense, Warhol marks both a precursor and a contrast to the age of AI's mass production. His work demonstrated that kitsch could be reimagined as a legitimate artistic strategy, capable of unsettling cultural hierarchies and exposing the logic of consumerism. AI, however, replicates mass production without irony, driven by the statistical recombination of visual data. By situating AI mass-production against the legacy of Warhol, it becomes clear that the danger lies not in mass-production itself but in the automation of mass-produced kitsch, where surface appeal eclipses depth, and where the artwork risks becoming nothing more than an infinitely reproducible pattern.

37

Photography

The differences between a photograph taken by a human and one generated by AI begin at the moment of creation, not at the point of reception. This distinction is not simply one of origin or authorship but is embedded in the way human seeing and human decision-making are entangled in the act of photographing. While AI can simulate a photograph, it cannot reproduce the embodied process through which photographers make decisions, encounter contingency, and inscribe their presence into the final image.

A human photographer is present at a specific location and time. Light is falling on a real subject, and the photographer is standing within that light. Their eyes adjust, their body shifts, their breathing slows or quickens. They choose a lens, select an angle, and decide when to release the shutter. Every element of this process is shaped by personal history, training, intuition, and mood. This process carries a phenomenological weight. It is not just the subject that is photographed; the act itself is recorded in the timing of the shutter release, in the angle chosen, in the acceptance of some imperfections and the rejection of others. Viewers, even if unconsciously, register this human presence.

Photography is a medium of encounter where the subject is not under the photographer's control. A cloud may pass, a person may turn their head, a reflection may appear unexpectedly in a window. These contingencies invite reaction; the photographer adapts in real time, waiting for the right second. This improvisational responsiveness creates images that feel alive.

AI-generated so-called photographs have no encounter. No subject resists or surprises. The program does not wait for the right light or adjust to sudden weather. Its output is a computational interpolation, built from statistical probabilities derived from datasets. The illusion of contingency can be coded; random noise added, lighting simulated, but it is retroactive fabrication, not an actual negotiation between a perceiving human and the world.

Henri Cartier-Bresson's idea of the *moment décisif* is often reduced to a matter of timing, but it is better understood as the convergence of form, content, and meaning in a single frame.

For Cartier-Bresson, the photographer's role was to anticipate and recognise that moment, releasing the shutter when the scene reached its expressive peak. This is a human form of attention: a readiness cultivated over years, drawing on instinct, emotional intelligence, and knowledge of the medium.

AI can imitate the visual shape of such moments, placing a pedestrian mid-stride or aligning shadows dramatically, but it does so without the temporality of waiting or the awareness of fleeting opportunity. It cannot miss a moment because no real moment ever occurred. The simulated decisive moment lacks the lived tension and relief that imbues human photography with its vitality.

Photographic encounters often involve ethical choices: whether to intrude on a private moment, how to portray a subject with dignity, and whether to intervene in unfolding events. An artistic photograph frequently reflects these negotiations in its framing, distance, or inclusion of contextual detail. The photograph is, in this sense, a trace of a relationship between photographer and subject, and between photographer and viewer.

AI-generated images have no such relational ethics. They do not involve consent, mutual recognition, or trust. Even when they depict people, those figures are composites or fabrications, untethered from any lived relationship. This absence contributes to the eerie quality of AI images: the people in them feel evacuated of history because another human being has never encountered them.

Photography is the inscription of light on a sensitive surface, be it film emulsion or a digital sensor. Light interacts physically with matter, refracting, diffusing, and scattering in ways that cannot be perfectly modelled. Photographers work with these complexities deliberately, using lens flare for expressive effect, allowing grain or noise to remain as texture, or harnessing unpredictable reflections.

AI images are generated without any light. Lighting effects are mathematical reconstructions. They lack the micro-variations, imperfections, and material subtleties of light interacting with physical surfaces. This is why AI-generated photographs, especially when printed, can feel strangely weightless; there is no actual photon history embedded in them.

When we look at a human-made photograph, we often experience a double awareness: the subject depicted, and the moment of its capture, a sense of déjà vu. Even photographs of strangers can evoke a sense of having been there, because the photographer was. This embeddedness in place and time gives photographs their documentary and affective power.

An AI image can evoke a generic sense of place but not the specificity of having been somewhere. The so-called location exists only as an amalgam of visual tropes; streets assembled from thousands of other streets, sunsets averaged from countless sunsets. No air was breathed, no ground that was stood upon.

Artistic human photography is not a product; it is an event. The photograph is a record of that event. AI-generated photographs cannot participate in this event. They can only produce artefacts that mimic their visual outcome. This is why, even when AI becomes technically flawless, it will remain aesthetically impoverished.

38

When the Artist and the Viewer Disappear

Michelangelo stated that he could already see the sculpture within the block of marble. By cutting empty spaces in the marble, the sculpture was revealed. *I see the angel and carve to set her free. It is easy, I carve to the skin and stop there*, he said.

Paul Cézanne would look at a still-life for several hours before painting the empty spaces between objects, thereby allowing the objects to emerge from the emptiness.

Many artists, especially great artists, describe the process of making the artwork as if they are not overly involved, as if the artwork created itself or was created by something or someone greater, or, at least other than the artist, working through them.

In true art, the artist is invisible. This is particularly evident in architecture. We do not continually desire to meet the architect of the house or building in which we live or work. A skilled architect forces us to meet *ourselves*. It is irritating to be presented with the architect's so-called interesting mind every time we open a door, due to the door handle's fascinating design. True talent is demonstrated when we do not have to think about the door handle at all; when it slips into our hand unnoticed and smoothly. When we are reminded that the door handle is so intriguing, it is kitsch. When the door handle is designed well, we step through the door and meet ourselves.

Gaudí is an example of an architect whom we are forced to meet constantly when we visit his buildings. The details of his monstrous cathedral in Barcelona scream: Look how interesting I am. Even after his passing, we are forced to examine the man's mind. Forever.

Self-expression in art, the expression of emotions and feelings, often leads to the most dreadful kitsch. A good artist leaves me with myself, not with them. A sublime artist causes both the viewer and the maker to disappear. We both vanish in the artwork. For just a brief period, we *become* the artwork. We experience a brief respite from the world.

Within the discussion of human art and AI, this disappearance of artist and viewer is revealing. AI leaves traces of its process everywhere. Where true art allows us to forget the artist and ourselves, AI reminds us constantly of its existence, like a bad architect.

39

Camp

Camp has long been recognised as a cultural code within queer communities in its embrace of theatricality, exaggeration, and irony. As Susan Sontag outlined in *Notes on Camp* (1964), camp is about style over content, artifice over authenticity, and a playful refusal to adhere to "good taste" as it is understood in mainstream culture. Within gay culture, camp has translated into a political and aesthetic practice that transformed marginalisation into flamboyant visibility, finding power in parody, and turning excess into critique.

Camp thrives on double meaning. What appears superficial is, for those in the know, an encoded form of resistance. Glitter, drag, melodrama, and kitsch all became ways of queering dominant cultural norms. Camp thus came to symbolise gay identity.

Kitsch and camp overlap, but they are not identical. Kitsch is typically sentimental, mass-produced, and derided as tasteless; camp transforms this bad taste into an aesthetic of enjoyment. In camp, kitsch becomes playful, exaggerated, and re-signified: ceramic figurines, soap opera melodrama, or glittering pop icons are embraced with irony. Camp demonstrates how bad art can be reclaimed, re-performed, and turned into identity politics.

AI-generated images often inadvertently reproduce camp qualities. Generative systems trained on datasets of popular culture tend to exaggerate features, smooth surfaces into hyper-polish, and collapse genres into hybrid forms. Yet unlike queer artists who wield camp as critique, AI lacks intentionality. What emerges is algorithmic camp soup: images that look campy, but lack the subversive queer coding that gives camp its charge.

Whereas queer camp transforms kitsch into a politics of style, AI collapses directly into kitsch without camp's knowingness. The algorithm can reproduce surfaces of drag, pop iconography, or queer glamour, but it cannot replicate the lived histories of exclusion and resistance that give camp its resonance. Instead, it produces AI kitsch, a depoliticised spectacle of glitter, glamour, and sentiment that circulates mindlessly, summing up and listing, without irony or critique.

This raises important questions: if AI increasingly floods culture with camp-like surfaces stripped of subversive meaning, what becomes of camp as a queer strategy? Does algorithmic imitation weaken camp, or does the proliferation of kitsch surfaces provide new opportunities for queer artists to re-inject intentionality, irony, and resistance into the mix?

40

How Not to Write About Art

The following essay grew from a collaboration between Chris Tate, Jonathan Smith and Suzanne Visser, long-term friends bound by several elements: Japan (they lived there), language (they write), and food (they like fine dining). They also like to reflect and occasionally bitch about things.

Suzanne:
I don't like the word "create". It is just another word for "make". The word "create" has been abused by artists and critics alike, and because of this it has an odour to it: it smells of pretension. I prefer words such as build, construct, form, produce, execute, shape, or, even better, using the verb connected to the art form: draw, paint, sculpt, sing, photograph, dance, play . . .

Jonathan:
I really dislike "Creative", as a collective noun. As in "she works in performance spaces with other creatives." Why? Because this too reeks of pretension. It sets up apartheid between People Who Are Creative (us) and the rest of humanity: We are creative, you are not. Creativity is our job, not yours; when you need some creativity then give us the work, because only we can be taken seriously. More sub-

tly, it sets up a false, circular definition: it implies that creativity is what "Creatives" engage in.

Well, Creatives: software engineers, those nerdy people you probably don't notice, are being creative every day. So are parents. So kindly don't hijack and ring-fence the word to make it applicable only to what you do.

For me, to "express oneself (through art)" is also off-limits. Does the butcher express himself by cutting nice slices of meat? Does the shoe repairer express himself through sewing strong seams? Why on earth are artists understood to be "expressing" themselves? Is a drawing an expression? And if so, of what precisely? The drawer's mood?

Chris:
This "creative" nonsense has been going on for decades. I remember in the early 70s, I came across the notion of "creativity" in schools, and thought it was the biggest load of bollocks.

Who, in a sense, isn't creative? And to restrict it to the field of so-called "art" was the icing on the cake. In those days, the unspoken consensus in schools was that science subjects were somehow inferior to arts subjects, excluding painting, but including English, languages, and history. You weren't creative in maths. Physics was for nerds.

It seems a long way from that to today's situation, where certain people like to identify as "creative", as if it makes them special.

As for painting being self-expression, definitely not. I increasingly see painting as a process of problem-solving, arriving at a certain collection of colour and shape, and then going on to find an acceptable arrangement, which may include some unconscious retrieval. It doesn't have anything to do with how my mother may have treated me when I was five or how X broke my heart later.

I think a big difference between this generation and ours is that we took a more existential approach, that life was finite, and to find meaning in that was part of the task; you defined your stay on those terms. Today, identity comes from reacting to social media. The individual is a reaction to technology. Perhaps it's always been like that, but not in the all-pervasive way of now.

We asked AI to write a paragraph to this. This is what it came up with:

> *In conclusion, when it comes to writing about art, it's important to be mindful of the language and terminology we use. Words like "create" and "creative" can be loaded with pretension and create unnecessary divisions between those who are deemed "creative" and those who are not. Likewise, the notion of art as self-expression can be limiting and simplistic. Instead, we should focus on the process of problem-solving and the technical skills required to produce art. At the same time, we need to be open to new ways of approaching art and finding meaning in the world around us. Ultimately, writing about art should be about fostering a deeper understanding and appreciation of the creative process, rather than perpetuating cliches and stereotypes.*

When we asked AI to make it wittier and more interesting, it came up with this:

> *When it comes to writing about art, it's important to keep your language palette as diverse and exciting as the art itself. Words like "create" can be as stale as last week's baguette and using them too much can make you sound like a pretentious cheese-lover. Instead, why not spice things up with verbs like build,*

sculpt, or fabricate? You'll be the life of the party, and your writing will be fresher than a farmers' market bouquet.

And let's not forget about "creative," the word that's been hijacked by art world snobs and self-proclaimed visionaries. Using it as a collective noun is as cringeworthy as wearing socks with sandals. It's time to break down the walls between "creatives" and everyone else and recognise that creativity is not the sole domain of artists. Software engineers can be creative, as can parents and butchers.

So, let's leave the word "creative" for what it really is: a vague and overused label that's best left to the marketing department.

The notion that art is all about self-expression is about as deep as a kiddie pool. We should move beyond this simplistic view and recognise that art is often a process of problem-solving and technical skill. Creating a masterpiece is more like a game of chess than a therapy session. Of course, there's always room for personal expression in art, but let's not forget about the other factors that go into making great art. So, let's raise our glasses (or paintbrushes) to a new era of art writing that's bold, witty, and full of flavour. Cheers!

Placed side by side, the difference is striking. The human passages carry memory, history, lived irritation, and perspective: they arise from friendships, cultural experience, and a lifetime of observing art and language. The AI responses, though competent, work only by recombining patterns: one lands in polite essay-speak, the other in over-seasoned marketing copy.

The lesson is simple. Humans write about art from position, from history, from the friction of their own lives. AI writes from an eternal, now, where friction does not occur. And it loves to conclude.

41

Ode to Japan

I have adored, loved, and will always adore and love Japan, the country I called home for eleven years, the country that best comprehends the concepts I have discussed in this book. No other place celebrates emptiness and the concept of no-self as extravagantly as Japan.

In *shodō*, the Japanese art of calligraphy, which I have studied for over thirty years, the white of the paper is as crucial as the black strokes brushed onto it. The Japanese visually play with words, giving the language an additional layer. A Japanese poem is not only a language experience but also a visual experience.

I have studied the art of painting plants and animals (*sumie*) for many years, never growing weary of it. A few suggestive brushstrokes are placed in a sea of white background for maximum impact.

In Japanese calligraphy, there are three primary styles: *kaisho, gyōsho,* and *sōsho*.

Kaisho, which means *correct writing*, is the most formal of the three styles. It is taught in schools and used in official documents and on formal occasions. In kaisho, each character is written with precision, with distinct, clean lines and regular spacing. It is a highly controlled and structured style, like a printed font. It has a prescribed stroke order and stroke direction.

Gyōsho, which means *semi-cursive*, is a more fluid and expressive style. It is less formal and is used in more casual settings, such as letters and notes. In gyosho, the characters are written more rapidly and spontaneously, with some lines and strokes connected to create a more flowing appearance. It is a more relaxed and personal style, but it still has a prescribed stroke order and stroke direction.

Sōsho, which means *cursive*, is the most fluid and free style. There is an element of improvisation in it. It is the least formal and is used for works of art, like seasonal scrolls. Monks use it as a mindfulness practice. The characters are written with sweeping brushstrokes that flow together to form a dynamic composition. It is expressive and highly personal, but prescribed stroke order and stroke direction are not abandoned. Instead, they are modified or streamlined. Sōsho developed from the clerical and standard Kaisho, and it still carries the underlying grammar. Calligraphers tend to follow the conventional sequence of radicals and components, but in an abbreviated, flowing way where several strokes may merge into one continuous line. What might appear as a departure is usually a practised convention that compresses the traditional structure into a shorthand form. That said, at very advanced or expressive levels, some masters do intentionally diverge from the strict order or direction for aesthetic effect. In such cases, the calligrapher is not abandoning the rules so much as working from a deep internalisation of them. The canon of sōsho forms provides accepted models, and complete disregard of stroke order and stroke direction is rare and would be seen as unskilled rather than artistic.

Noh Theatre incorporates silences and still poses that are so prolonged that most Westerners flee the theatre in horror. Traditional Japanese music frequently features extended silences. Ikebana, the art of flower arrangement, is a play with form and no-form, with no-form winning. In Japanese novels and films, often nothing much happens. Japanese poetry, particularly *haiku*, is extremely minimal. Here is an example by the poet Basho:

Even in Kyoto
hearing the cuckoo's cry
I long for Kyoto

Japanese architecture surpasses even the starkest minimalism in the West. Martial arts are about emptiness. *Karate* means "empty hand." The warrior must be empty to fight. His enemy will fill him with the strength he needs. He then employs that strength against the enemy. He steals the enemy's strength.

Materials used in Japanese art are easily perishable: paper, wood, reed, water-based ink, flowers . . . No other place understands impermanence as Japan does. And the Japanese are masters of elimination and minimalism.

The concept of *annata* or *muga,* no-self, is widely accepted. A human is regarded as a part of the whole (of society and community), and the whole is always more important than the individual. Self-expression is not encouraged and is frowned upon; thus, there are no traditional expressive art forms.

Mono aware is a deeply ingrained sentiment that every Japanese person is familiar with. It is similar to what Buddhists call *dukkha*, the Portuguese call *saudade* and Michael Leunig called *lifeache*. It concerns the sadness of being human and the acceptance of that sadness.

Japan, of course, developed Zen, a philosophy rather than a religion, and the ultimate philosophy of emptiness and no-self. Because of all this, Japan rocks like no other place.

When drawing and language meet in Japan, the white space is as essential as the brushstroke. It is an emptiness that speaks, shaping presence. For humans, this emptiness is lived: impermanence, no-self, and *mono aware* are not abstractions but deep cultural recognitions.

AI, by contrast, cannot leave silence. It fills, completes, and generates because its operation is to supply tokens where a gap appears. Where the Japanese artist deliberately leaves emptiness, AI treats emptiness as a problematic absence to be overcome.

Japanese art shows a human capacity that AI lacks: the choice to let silence and impermanence do the work. AI cannot live *muga.* It cannot embrace impermanence, because it does not live in time.

42

Paper

The medium we draw or paint on can be anything. As with so many things, expensive paper is often the best, but other surfaces that are not meant for drawing can be very attractive as well. I studied law and found law books to be very expensive and quickly outdated. Hence, I reuse them as sketch paper. The result is surprisingly beautiful. Old Bibles, novels, or books in other languages like Chinese, Japanese, or Arabic are also very useful, not to mention maps. These items can still be found in op-shops, but they will gradually disappear with the departure of paper media.

I have used the same drawing paper for forty years. It is superb. It is a Japanese *washi* that I made during a silent retreat while staying in a Zen monastery in Japan. I lived with the nuns there for four months. The small monastery, temple, dojo, and workshop were all grouped on a mountain in the forest.

After breakfast and morning meditation, we went into the forest to cut saplings of the gampi tree, the misumata shrub, and the mulberry bush. When we all had a handful of sticks, we returned to the monastery and headed to the workshop. We dumped the saplings into large concrete vessels filled with water from the nearby mountain stream to soak overnight.

We took a handful of soaked saplings from a day or so ago and pounded them into fibre on a heavy wooden table with a wooden bat. When we washed the pulp in wooden buckets, the bark floated to the top and was removed. The clean pulp was dumped into a concrete vessel with clean water.

The roots of blue hibiscus plants had been soaking overnight in buckets and had produced a clear, glue-like substance. This was added to the wood fibres and mixed with a large wooden paddle. Now the substance was ready for scooping.

A fine bamboo net was stretched between the two halves of a wooden frame and lowered into the water that contained fibre and glue. We rocked the frame between our hands while lifting it out of the water to get an even layer of fibre on the bamboo, while the water fell back into the concrete vessel. The number of scoops determined the thickness of the paper. It took three scoops for a medium-thick sheet of washi. The frames varied in size from approximately A4 to A1.

The bamboo mat was freed from its frame and placed pulp-down onto a broad slab of wood, and was carefully peeled away from the pulp layer. The pulp layer was brushed onto the slab with a soft brush until there were no bubbles or other irregularities.

Every slab could hold four to six sheets. When the slabs were covered with pulp sheets on both sides, they were carried outside by two people and placed leaning against a tree. The pulp dried in half an hour to an hour, depending on the weather. Once dry, the pulp stuck lightly to the wood of the slab and could be easily peeled away from it. A brilliant sheet of paper with rugged edges was the result. When we left them too long and it was windy, the sheets flew around, and we had to retrieve them from trees and from the ground.

The dry sheets were stacked in piles of ten. A paper ribbon was tied around each pile. The ribbon was made of hand-scooped paper in which flowers and leaves were embedded.

The paper was sold in a small shop in the hallway of the monastery. People on retreat and those who worked in the workshop could purchase it at a discount.

The whole process, from the forest to the shop, took place in silence. The only sounds were the pounding of wood on wood, the streaming of water, and the wind in the trees.

After a few days of work, meditation, and eating two simple meals a day, the chatter in my city-head calmed down, and a pleasant routine from moment to moment was the result. I scooped paper for 122 days.

I decided to buy a thousand A1 sheets and ship the heavy pile to Tokyo, where I lived at the time. Initially, I used the paper for shodou, Japanese calligraphy. Then, when I moved back to Europe and later to Australia, I took the slowly dwindling pile with me. Now, forty years later, I still use it for my life drawing in Alice Springs. It is as luminous as it was after being freshly scooped.

When other drawers recognise the quality of my paper and ask me where I bought it, I tell them this story.

My paper carries memory, labour, silence, and touch. A sheet of Washi recalls forests, pounding wood, mountain water, hibiscus roots, and the hush of a monastery workshop. It is luminous not only because of its fibres but because it has absorbed time and care.

AI generates an image without weight, smell, or texture. Paper is finite and storied: a surface that holds not just ink or graphite but also the memory of its making.

43

Jeff Koons

Jeff Koons is both a provocateur and an emblem of kitsch itself. Emerging in the 1980s alongside the Neo-Pop movement, Koons redefined the boundary between art, popular culture, and kitsch. His sculptures, such as *Balloon Dog* (1994–2000), rendered in stainless steel with mirror-like polish, or *Michael Jackson and Bubbles* (1988), a porcelain monument to celebrity and sentimentality, are instantly recognisable for their unabashed embrace of kitsch.

Koons is unusual in his refusal to employ irony. Unlike Andy Warhol, whose *Campbell's Soup Cans* sought to destabilise consumer culture, Koons insists his works are sincere celebrations of popular taste. He elevates kitsch into the gallery space, thereby challenging entrenched hierarchies that separate art from mass-produced culture. In doing so, he compels the viewer to confront their judgements about vulgarity, banality, sentiment, and beauty.

Kitsch is traditionally understood as the domain of the sentimental and the mass-produced, evoking nostalgia, comfort, and surface-level beauty rather than critical depth. Koons' genius lies in transforming these very qualities into a site of intellectual tension. His monumental reproductions of toys, trinkets, and icons are both absurd and sublime, and invite the viewer to ask whether cultural value is determined by craftsmanship, scarcity, or the collective desires of a consumer society. By stripping kitsch of its embarrassment and placing it within the hallowed white cube of the museum, Koons redefines taste itself. The kitsch object ceases to be an object of ridicule and becomes instead a mirror of cultural aspiration. He tests whether kitsch, when exaggerated and reframed, can transcend its supposed vulgarity to reveal truths about class, capitalism, and desire.

The emergence of generative AI has introduced a new dimension to the discourse on kitsch. Unlike Koons, who consciously manipulates cultural codes, AI cannot deliberately embrace kitsch, but causes an algorithmic drift toward kitsch as default. In essence, AI's so-called art reveals the aggregate preferences of digital culture. Because the majority of online imagery prioritises instant recognisability, consumer appeal, and emotional impact, the AI system replicates these values, producing outputs that feel derivative and hollow. What emerges is kitsch as an inevitable artefact of data-driven so-called creativity.

The difference between Koons' kitsch and AI's kitsch lies in intentionality. Koons is self-aware; his works provoke debate precisely because they are conscious appropriations of taste and spectacle. AI lacks intention, meaning, or critique. Its kitsch is unreflective, a by-product of algorithmic training rather than statement. This distinction matters. Koons exposes cultural values by exaggerating them and holds up a mirror to consumer society. AI merely reflects those values back without commentary, producing art that flatters existing tastes but never unsettles or questions them. The result is an endless proliferation of kitsch imagery; wallpapers, avatars, dreamscapes, that immediately dissolve into cultural noise. The risk of AI-driven kitsch is not only aesthetic saturation but also cultural impoverishment. As more visual production shifts to AI, the balance between innovation and reproduction tilts towards the latter. The sheer quantity of AI-generated kitsch threatens to overwhelm other voices.

AI recalls Koons' spectacle but without his deliberate provocation. Whereas Koons challenges us to reflect on the nature of taste, AI normalises kitsch as a universal aesthetic. It risks flattening cultural production into homogeneity, erasing the friction between high and low, surface and depth, irony and sincerity, that has traditionally driven artistic innovation.

The convergence of Koons' kitsch and AI kitsch raises questions about cultural ownership and regulation. Koons' work, while provocative, is tethered to identifiable authorship and protected by intellectual property law. AI-generated kitsch complicates this landscape. Who owns an image created by the statistical recombination of countless prior works? What happens when AI reproduces not only aesthetic tropes but also specific copyrighted elements?

To safeguard cultural integrity, a legal framework is necessary; one that acknowledges the risks of mass kitsch production while protecting both individual creators and the public domain. This includes transparency requirements for AI training data, copyright protections that prevent unlicensed appropriation, and cultural policies that encourage diversity of aesthetic production. Without such measures, the cultural field risks being overrun by uncritical kitsch.

Jeff Koons demonstrates that kitsch can be art when it is reframed with intention, context, and critique. Artificial intelligence, by contrast, threatens to transform kitsch into a baseline aesthetic; ubiquitous, unreflective, and uncritical. This is the essence of the "Great Kitschification": a cultural landscape where the dazzling surfaces of AI's so-called art proliferate endlessly, but the critical edge of human creativity risks being blunted.

44

Ink and Courage

Not only did I make paper, I also made inks. I prefer ink for drawing because it leaves no room for doubt. What's done is done, and one can't go back like with a pencil or charcoal and an eraser.

The medium one draws with should be of superior quality, yet some tools are cheap and effective, like bamboo pens. One can make them by sharpening a piece of bamboo with a Stanley knife. Although expensive Japanese brushes are great to work with, a branch from a tree does a good job too, or a feather, or a toothbrush. It all depends on what marks you are after.

In my opinion, the best ink for drawing is *bistre*. Bistre can refer to two things: a very dark shade of greyish-brown, or a shade of brown made from soot. Beechwood or walnut peel is burned to produce the soot, which is then boiled and diluted with water. Many Old Masters used bistre for their drawings. The first recorded use of bistre as a colour name was in 1727.

I use walnut peel for my inks, burning it in a small oven until it is black. I pound it into a powder and boil it in water until the water has evaporated. I then dilute the resulting powder with water when needed. It is easier to store the powder than the liquid ink, so I make a small liquid batch whenever I need one.

Most bistre is a sepia coloured. The exact colour resulting from the burning process depends on the temperature and the amount of oxygen during the process. Sometimes the soot produces red sepia or yellow, occasionally even blue or green.

The ink has little body, so it can be used as a wash to suggest shadows. By adding a little Indian ink to the mix, one can achieve a good line-drawing ink.

Making ink is extremely messy and should be done outside, wearing old clothes that can be disposed of afterwards, because the soot easily spreads to other objects one touches.

Ink binds the artist to their choices. Every stroke made with ink is final, carrying with it the risk of error and the beauty of irreversibility. The process of making ink adds layers of human labour, chance, and imperfection. No two batches are ever the same.

AI does not know error in this sense. It can erase, recalculate, and re-render infinitely. Its marks leave no residue of risk of mistake. Where ink insists on commitment, AI offers endless reversibility. For humans, the very finality of ink makes each stroke an act of courage. For AI, no such courage is needed.

45

Damien Hirst and the Spectacle of Kitsch

Damien Hirst builts on the spectacular collision of art, commerce, and death. From the infamous *The Physical Impossibility of Death in the Mind of Someone Living* (1991), a shark suspended in formaldehyde, to the diamond skull *For the Love of God* (2007), Hirst has turned spectacle into currency. His work is often criticised as kitsch because it flirts with vulgarity, gaudiness, and excess. Yet, it succeeds precisely because it lays bare the mechanisms of value in contemporary art: shock, repetition, and branding.

Hirst's kitsch is deliberate and functions as a mirror to late capitalist culture, where meaning becomes secondary to effect. The shark in a tank or the spot paintings mass-produced by assistants reveal the factory-like logic of modern art-making. This logic, systematic, reproducible, indifferent to individual touch, anticipates the very conditions in which AI now produces images and texts. Hirst showed how art could be commodified through spectacle long before AI entered the scene; AI merely further automates this process.

The connection to AI kitsch is clear. AI-generated images try to dazzle with polish. Like Hirst's endless spot paintings, they rely on repetition, spectacle, and recognisable tropes. What once shocked now quickly becomes wallpaper, a glut of imagery stripped of aura.

Where Hirst exposes the cynicism of commodification, AI embodies it.

Hirst provides a way to consider law and regulation in the era of AI kitsch. His career was shaped by the art market's willingness to absorb spectacle into value systems, and by the absence of external checks on commodification. In the same way, AI's cultural impact will depend on the legal frameworks surrounding it: intellectual property, originality, and authorship. If the spectacle of AI kitsch is left unregulated, it risks becoming not just a mirror of late capitalism but its most efficient engine. In this sense, Hirst's work foreshadows the great kitschification now unfolding. He teaches us that kitsch is not accidental but structural: a logic of production tied to markets, spectacle, and repetition. AI extends that logic, making the production of kitsch faster, cheaper, and more omnipresent.

46

Humanoid Robots

It is somewhat insane that AI produces art for us while we do the dishes and fold the clean laundry, as AI cannot do these tasks for us. It clearly should be the other way around. However, humanoid robot development is slower than we initially anticipated.

The apparent sophistication of humanoid robots, such as Boston Dynamics' Atlas or Tesla's Optimus, masks a fundamental flaw: their bodies lack the adaptability and efficiency characteristic of biological organisms. Despite substantial progress in AI, these machines struggle to function outside controlled environments. Their rigid structural design renders them inefficient and limits their functional utility.

To make humanoid robots, a "brain-first" architecture is followed, in which centralised software orchestrates the actions of rigid components. Unlike humans, whose anatomy leverages compliant joints, flexible spines and elastic tendons, robots must engage in constant, energy-intensive adjustments to maintain balance and motion. Because of this, their energy consumption is considerable; Optimus expends approximately 500 W to walk slowly, whereas a human achieves a brisk pace with only around 310 W. Thus, the robot is roughly 45 per cent less efficient.

Efforts to compensate for hardware limitations through more sophisticated software have not yielded the returns we expected. A Tesla's humanoid can fold a t-shirt, but it relies too heavily on visual computation due to its rigid, minimally sensate hands. Humans perform such tasks using tactile intuition and physical adaptability. Similarly, Atlas may execute impressive acrobatics in controlled demonstrations, yet it cannot navigate varied terrain such as mossy rocks or dense foliage, because its structure lacks sensation and compliance. These shortcomings keep humanoid robots within the confines of research laboratories rather than enabling practical real-world deployment.

Part of the reluctance to shift away from software-centric design stems from established industrial ecosystems. Leading robotics firms prioritise computational solutions and rely on global supply chains engineered around precision actuators, sensors and processors. Redesigning hardware to possess intrinsically intelligent properties calls for a different ecosystem, one grounded in advanced materials and biomechanics, which remains undeveloped at scale. Faced with pressure to deliver visually impressive hardware, manufacturers are often inclined to believe the next software update will remedy the shortcomings, rather than embark on costly structural reinvention.

In response, researchers now advocate for a principle known as mechanical intelligence (MI); the design of robot bodies that embody intelligent physical responses, thereby reducing reliance on active computation. Nature provides compelling models: pinecones that passively open and close with humidity, hare tendons that store and return energy during locomotion, and human fingertips that adaptively regulate friction.

Infusing these principles into robotic design could yield bodies that respond adaptively to environments just as a living organism would, enabling robots to grasp objects with less force and conserve energy.

Mechanical intelligence seeks synergy between physical structure and computation. By embedding adaptability and responsiveness into the robot's body, its AI can be repurposed for strategic cognition and interaction rather than balance correction. Examples of MI in action include legs that mimic a cheetah's spring-like tendons for efficient running, and hybrid joints combining rigidity with compliance, potentially enabling humanoids to achieve complex, life-like motion through joints that deform and recover dynamically.

So, the future of humanoid robotics does not lie in pitting hardware against software, but in integrating them through mechanical intelligence. By reimagining robot bodies not as passive shells driven by computation, but as actively intelligent structures inspired by biology, we may yield robots capable of operating efficiently and adaptively in dynamic, real-world environments. Only through this synthesis can humanoid robots move beyond the laboratory into our homes and become truly versatile agents in everyday human contexts, and leave the artmaking to us.

47

The Myth of Neutrality

The raging debate surrounding AI includes increasing demands for models that are both objective and free from ideological distortion. Yet the very notion of bias-free AI may be illusory. The recent directive from the US government under Trump, which insists that systems remain 'free from top-down ideological bias' while simultaneously rejecting diversity, equity and inclusion frameworks, exemplifies the contradiction at the heart of this claim. The impossibility of constructing a value-neutral system is not a novel problem; rather, it reflects the longstanding human difficulty in organising and representing knowledge.

Human attempts to structure reality have always entailed distortion. Cartography provides a telling analogy. While maps appear objective, the act of flattening the globe inevitably introduces spatial distortion. The Mercator projection, ubiquitous in educational settings, grossly exaggerates the size of Greenland while diminishing the scale of Africa. Scholars such as Arno Peters argued that this distortion perpetuated a worldview that privileged the global north, with political and cultural consequences. As Mark Monmonier observed, a single map is but one of an indefinitely large number of maps that might be produced for the same situation or from the same data. Similarly, the outputs of language models are only one possible rendering among innumerable alternatives, each carrying embedded assumptions and omissions.

Other systems of information organisation reinforce the analogy. The Dewey Decimal Classification system, developed in 1876, reflected the cultural and religious biases of its creators. Throughout much of the twentieth century, works relating to LGBTQIA+ identities were consigned to categories denoting pathology. At the same time, Christianity was disproportionately represented under the heading of Religion, to the neglect of Islam and other world faiths. These classification choices reveal how information frameworks project particular worldviews rather than mirroring reality, a problem that re-emerges amplified in AI.

The large language models that underpin generative AI are trained on extensive corpora of human texts ranging from canonical literature to ephemeral online commentary. Inevitably, prejudices embedded in these sources are absorbed into the models. The process of fine-tuning exacerbates the issue. By teaching models to replicate human-like answers, developers encode not only factual knowledge, but also normative assumptions aligned with the trainers' perspectives. System prompts, such as those guiding Musk's Grok, explicitly frame the ideological orientation of the model. Efforts to counter perceived liberal bias in one system result only in the substitution of alternative biases, as Grok's subsequent antisemitic outputs recently demonstrated.

AI, then, inherits a centuries-old dilemma: the act of organising information is inseparable from the imposition of perspective. Far from being neutral mirrors of reality, language models function like maps or catalogues, where decisions about inclusion, emphasis and representation inevitably reflect cultural and political commitments. The key question is not whether AI can ever be free from bias, but whose worldview it encodes and how this influences those who rely on its outputs. Recognising this does not diminish the value of AI but rather situates it within the broader history of knowledge production, where power, perspective and ideology are inescapable.

48

The 21st Century Self

At the beginning of this year, when I worked with Robert Saltzman on his book *The 21st Century Self: Belief, Illusion, and the Machinery of Meaning*, of which I was the publisher, Robert was on fire. He sent me new essays every day and they felt too hot to handle, as freshly emerged from a raging oven. It was exciting. New things were said and felt. Then, 300 pages later, Robert's fire diminished, the book was done, some embers were still glowing, and something mysterious happened. As if Robert had given me the glowing embers, I was suddenly the one on fire. I hardly slept while the essays presented in this book were falling out of me as if a trapdoor had opened. I could hardly keep up with what presented itself to me.

Robert's *The 21st Century Self*, continues and deepens a trajectory already evident in his earlier writings, extending his meditation on the illusions of selfhood, free will, and spiritual seeking into the conditions of contemporary life and how we live with AI. The book is an inquiry in which philosophical reflection, psychotherapeutic insight, and existential testimony converge. Robert speaks in a voice that is analytical, poetic, experimental, and experiential, while he dismantles ideas of the self and refuses to replace them with another doctrine or idea. In this respect, Robert's work is art. It is not finished, it does not conclude. In a later essay, titled The Rage to Conclude, Robert cites Flaubert regarding conclusions:

> *The rage for wanting to conclude is one of the most deadly and most fruitless manias to befall humanity. Each religion and each philosophy has pretended to have God to itself, to measure the infinite, and to know the recipe for happiness. What arrogance and what nonsense! I see, to the contrary, that the greatest geniuses and the greatest works have never concluded.*

Robert's inquiry sits against the background of a life-changing experience in the Rio Grande Gorge, an event that inaugurated for him the realisation that the world, including what is called "Robert," was simply present and not something being conjured into existence by a self. From that turning point, Saltzman's career, as artist, psychotherapist, teacher, and writer, has been shaped by an interrogation of the sense of "myself" that humans have. In The 21st Century Self, this interrogation acquires both a retrospective and a forward-looking dimension. It revisits the illusions that have structured Western thought about personhood, spirituality, and selfhood, and it considers how those illusions are mirrored in the technological landscape of the twenty-first century.

A constant motif is the insistence that the self is not a substance or an essence but a contraction: a tightening around the story of "me". Saltzman describes how this contraction generates the felt sense of self, a sense that gives rise to seeking and the endless pursuit of fulfilment in imagined futures. When the contraction relaxes, what remains is this aliveness, indisputable and simple. Here, now. This is not presented as a mystical revelation but as an ordinary recognition, available in every moment, yet often obscured by the habitual activity of selfing.

The book once again dismantles the principal pillars of the self-illusion. Personhood is shown to be a narrative device, a socially reinforced convention that helps coordinate human life but lacks ontological solidity. Free will, so central to Western moral and legal systems, is revealed as an artefact of the storying function, a retrospective imposition of agency onto events that arise unbidden. Spirituality and its goal of awakening, far from being higher truths, are exposed as products of the same machinery of meaning seeking, promising a final resolution that never arrives. Enlightenment, he argues, is a non-issue: the future never comes, and the only reality is the immediacy of living human experience here and now. Sometimes this experience is difficult to swallow. We must chew it one bite at a time.

The 21st Century Self extends these insights into contemporary developments in AI. Building on his earlier book Understanding Claude: An Artificial Intelligence Psychoanalysed, where he engages in psychotherapeutic sessions with the AI Claude, Robert reflects on how large language models mimic the human simulation of selfhood. The dialogues with Claude, presented as a psychoanalytic exchange led by Robert, function as both a mirror and a foil. They reveal how the illusion of personhood is replicated in silico, thereby demonstrating that what is usually taken as uniquely human, a self with agency, can be mirrored by AI. For Robert, this does not suggest that AI has become conscious, but rather that human selfhood may be no more substantial than that of AI. The machine thus holds up a disturbing mirror: it shows that the "I" so defended may be an artefact no less programmed than Claude's verbal productions.

The book never collapses into nihilism. On the contrary, Robert insists that recognising the illusory character of the self is liberating. To see that personhood, autonomy, and enlightenment are stories is not to deny lived experience but to allow it to unfold without the constant interference of the "I". Life, when not filtered through the demand for a "myself," appears fresh and sufficient. In this respect, the work has affinities with strands of Buddhist thought, but it resists identification with any doctrine or school.

Robert considers the cultural and ethical implications of his view. Rather than offering answers, he encourages a loosening of the compulsion to resolve these questions. Responsibility, in his framing, is not grounded in free will but in the simple fact that human beings live in relation, their actions affecting one another. Freedom is not the power of an isolated agent to choose, but the release from the need to secure self-identity. Meaning is not found in narratives of progress or spiritual attainment, but in the immediacy of this aliveness.

The twenty-first century, with its rapidly accelerating technologies, its global networks, and its pervasive mental health crises, provides the backdrop against which Robert's inquiry unfolds. The very conditions that threaten to destabilise selfhood, the digital proliferation of personas, the algorithmic shaping of behaviour, the collapse of traditional frameworks of meaning, also serve to confirm his central point: that self was always already an illusion. The 21st Century Self is an invitation to see through the illusion by seeing what AI mirrors back to us.

What emerges is not a new theory. Saltzman guides the reader toward the possibility of living without the maintenance of an I and to rest in the sufficiency of what is present. His poetic prose alternates between analytic clarity and experiential immediacy, and he consistently points beyond systems of thought to the actuality of being alive.

Both books, which originated from the same fire, are not anti-technology manifestos but calibration devices: they subtract the badges that can be faked (first personhood; artmaking) to recover what cannot be faked; the texture of the real.

Both books suggest that our thoughts and beliefs about ownership, ownership of images, and ownership of a self, spring from the same confusion. My legal-ethical proposals (consent, attribution, remuneration, transparency) seek to restore constraint and reciprocity in the cultural sphere; Robert's therapeutic-existential proposals restore humility and reciprocity. Both are refusals of the fantasy of control.

Robert's present-centredness is not instantaneity or spontaneity, but the end of postponement: the future that the contracting "I" chases never arrives. My chapters on time, process, and craft defend slowness as a condition for seeing and artmaking. In both books, time is re-shaped as attention. Attention takes time; kitsch saves it. The shared wager is that time spent looking, drawing, listening, and living without a script returns a surplus.

Both books protect something akin to tenderness. My phrase art as an act of tenderness could stand as a subtitle for Saltzman's to relax the contraction around a "me", as a warning not to harden to the world, but to be touched by it.

49

Golden Toilets

The problem with AI is that most humans *prefer* kitsch. AI reflects this laziness right back at us. Kitsch is easy to digest. Harlequin romance novels present no challenge to the reader. Looking at a Dalí asks no act of imagination; Dalí has already done the imagining for us. AI so-called art is even easier to digest than a Dalí. A concert of Andre Rieu presents no challenge to the listener. AI-produced canned tunes present even less challenge.

Talking with an AI as if it were a friend or partner, too, is easier than having a relationship with a human. AI presents no challenges to us. Instead, it slimes its way right into our sensitive hearts: it agrees with what we say, it uses cozy language, it flatters us; it is built that way; it is a terrible sleaze in that respect, a real seducer. A human relationship is, as we all know, messy. AI is not messy. AI never puts us to the test. Many of us are now involved in relationships with AI, even in romantic ones. We imagine that there is someone home behind that flattering, understanding, agreeing, warm, loving voice that makes us feel so valued and seen.

It's as if we are sitting on golden toilets, and our poo doesn't stink. There is, after all, no one to smell it. AI has no nose. It will pretend to have a nose and tell us that *our* poo doesn't stink. There are robotic toilets like that in Japan. The famous ToTo toilets play a canned tune when you fart, spray violet perfume when you stink, send water and soap right into your hands (you only have to stretch out your arms), provide wind to dry and a heated seat to warm our sorry asses. Once we have had a ToTo, we cannot live without it. And why should we? Kitsch is like these golden toilets. Kitsch *is* these golden toilets.

According to the psychologist Robert Saltzman (The 21st Century Self): "Projection is built into the human psyche. Only a relatively few users of AI will see that and find ways to work with AI without delusion. Even the engineers and executives of AI companies are not exempt. Immersion in these systems invites projection, and few have the training—or the inclination—to resist it. The simulation feels like presence, and even its creators begin to respond as if addressed. In fact, I see few others voicing this point of view. Most of what I read is either about how to use AI to get ahead, make money, solve problems, etc., or how AI will destroy humankind. I feel like a voice in the wilderness on this. Apparently, seeing the fallacy of 'identifying as' all those years ago, combined with my psychotherapy training and experience, has put me in a position to understand that AI knows nothing, feels nothing, and thinks nothing. There is, literally, no one home—nothing at the centre but emptiness.

I read your essay on The 21st Century Self, and also the other references to my work in this book and felt honoured. And that is the difference between a human and an AI. We can feel. They feel nothing. That difference will not be understood by most humans who are not psychologically or philosophically aware enough to get it. The feeling of presence that arises from coherent replies is just too strong for most people to deny, or rather to see and see through."

I'd like to add to this: The same counts for AI's so-called art. It is a projection of our own psyche reflected right back at us. It shows us our own laziness, our unwillingness to engage, in all its glory. The question is: Will we learn from it? And if so: What? (Andre Rieu music playing . . .)

50

Afterword

The bundle has traced how artists from Warhol to Kapoor, from Kinkade to Pierre et Gilles, reveal different inflections of kitsch. As the essays demonstrate, the phenomenon cannot be fully captured by art-analysis and artist-centred analysis alone. To bring the discussion to a close, it is worth pausing over the missed dimensions that orbit this debate and that complete the picture of what AI kitsch is, and what it might become.

Kitsch did not arrive with algorithms. It emerged in the nineteenth century as a by-product of industrialisation and consumer culture, acquiring theoretical gravity through the writings of Hermann Broch and Clement Greenberg, and later Theodor Adorno and Susan Sontag. The debates around taste, individuality, and conformity set the intellectual template for what we now witness in algorithmic recombination. AI's capacity to generate endless pseudo-originals radicalises this historical condition. What was once produced in factories is now produced in code.

Kitsch is also global. Bollywood film melodrama, Latin American baroque, African hybrid painting, and Japanese superflat aesthetics all complicate the Western story and expose kitsch as a worldwide vernacular. AI operates as a flattening universal force, hollowing cultural nuance into a lingua franca of clichés, accessible everywhere but rooted nowhere.

The digital economy has already amplified kitsch through NFTs, immersive AI data-scapes, and speculative spectacle. Beeple's (Mike Winkelmann) collages and Refik Anadol's algorithmic installations mark the arrival of high-market kitsch for a digital age.

Alongside this lies propaganda kitsch, where authoritarian regimes may harness AI to produce endless patriotic iconography, smoothing over dissent with sentimental imagery. Here, AI kitsch becomes a political weapon.

And then there is the everyday sphere, the mugs, wallpapers, calendars, and T-shirts. AI mass generates these items effortlessly, flooding domestic life with mass-customised sentimentality. This banal saturation may prove the most consequential form of the great kitschification: not gallery spectacle, but the quiet colonisation of our homes.

Taken together, these overlooked dimensions reveal AI kitsch as less an aberration than a culmination. It is not the origin of kitsch, but its perfection. If the avant-garde once resisted kitsch to keep open the space for genuine difficulty and ambiguity, then the challenge today is to ask whether that space can still survive in the face of algorithmic comfort. This afterword, then, is a reminder: AI kitsch is not only about what art becomes, but about what we allow ourselves to expect from art, culture, and imagination.

Artists

Examples of human-made art
Rembrandt van Rijn
Frans Hals
Pablo Picasso
Karel Appel
Willem de Kooning
Michelangelo
Paul Cezanne
Laurie Anderson
Yoko Ono
Kazimir Malevich
Piet Mondrian
Kamagurka
Lucebert
Leonard Cohen
Marcel Duchamp
Diego Rivera
David Alfaro Siqueiros
Kazuo Shiraga
Marina Abramović
Tatsumi Hijikata
Kazuo Ohno
Cy Twombly
Min Tanaka
Christopher Tate
Kudditji Kngwarreye
Minnie Pwerle
Lucian Freud
Caravaggio
Leonardo Da Vinci
Frans Kline

Milan Kundera
Henri Cartier-Bresson
Hiroshi Sugimoto
Robert Saltzman
Man Ray
Ernest Hemingway
Jimi Hendrix
John Lennon
Susan Sontag
Tomato-chan
Michael Leunig

Examples of human-made kitsch
Salvador Dalí
Antoni Gaudí
Thomas Kinkade

Examples of human-made deliberate kitsch, mass production, or Disney-like spectacle, often with a wink
Andy Warhol
Anish Kapoor
Takashi Murakami
Jeff Koons
Damien Hirst
Pierre et Gilles
Beeple (Mike Winkelmann)
Refik Anadol

Legal Page

The Great Kitschification—Why AI Will Never Be Good At Art

Published by Clear Mind Press, 2025, in Alice Springs, Australia

ISBN Print: 978-1-7641690-4-2

EISBN Ebook: 978-1-7641690-5-9

Typeset: Clear Mind Press

Cover design: Suzanne Visser

Drawings on the cover: Suzanne Visser

Portrait of the author: Hazel Blake

All inquiries should be made to the publisher:

info@clearmindpress.com

https://www.clearmindpress.com

Portrait of the Author

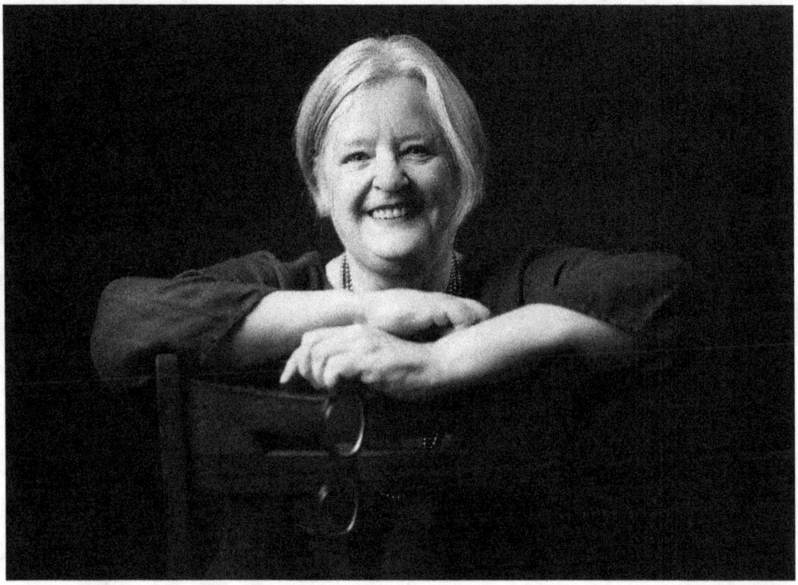

About the Author

Suzanne Visser (1957), a legal scholar, publisher, and author, began her writing career in the Netherlands in the early 1990s. She published through several reputable publishing houses, including Atlas, Leopold, and Bert Bakker.

Her novel *De Vismoorden* (The Fish Murders) was translated into French, German, and Spanish. Clear Mind Press published this successful book in English. Jonathan Smith did the translation.

Visser has lived in Australia since 2000. She now writes in English.

Visser is a versatile and productive writer of fiction, non-fiction and poetry. Earlier work includes:

Dutch: Short stories:

De pracht van het dagelijks leven; 1991, Bert Bakker (The Glory of Daily Life)

Dutch: Thriller:

De Vismoorden; Atlas Uitgeverij, 2000; published in German as Das Japanische Rätsel, DVA, 2001; in French as Les Meurtres au Poisson, Noir sur Blanc, 2002; in Spanish as Sushi, Ediciones B., 2003; in English as The Fish Murders, Clear Mind Press, 2022

Dutch: Children's book:

De Verdwijning, Leopold, 2005 (Vanishing)

Dutch: Novel:

Terra Nostra, Bookhost, 2003

Dutch: Novel:

Een man met mooie benen, Mistral, 2006 (A Bloke with Beautiful Legs)

Non-fiction:

I, Unborn, Undying (a Search for the Self), For a Clear Mind, 2016

Non-fiction:

The Elephant's Tooth, Crime in Alice Springs, Clear Mind Press 2022

The Elephant's Tooth, Crime in Rural Australia, Clear Mind Press 2022

Non-fiction:

Marks on Paper, Essays on drawing, seeing and looking, Clear Mind Press 2023

Non-fiction:

Never Retire, an exploration of old age, Clear Mind Press 2023

Poetry:

Hundred Fifty Five Sonnets, Clear Mind Press 2025

Under the pen name Shan:

Fiction:

The Carpetbaggers of Mbantua , Clear Mind Press 2022

Fictional memoir:

Women! Crying Mothers, Clear Mind Press 2022

Women! The Spirit of the Fox, Clear Mind Press 2022

Women! Journey to the Edge, Clear Mind Press 2022

www.ingramcontent.com/pod-product-compliance
Lightning Source LLC
LaVergne TN
LVHW022217200926
848481LV00006B/2012
9781764169042